Obtaining and Characterization of Phosphate Layers on the Surface of Steels for Reinforced Concrete

Petru LAZAR, Alin-Marian CAZAC, Costica BEJINARIU

Gheorghe Asachi Technical University of Iasi, Romania
Materials Science and Engineering Faculty

Published by **Materials Research Forum LLC**
Millersville, PA 17551, USA

Published as part of the book series
Materials Research Foundations
Volume 183 (2025)
ISSN 2471-8890 (Print)
ISSN 2471-8904 (Online)

Print ISBN 978-1-64490-380-3
ePDF ISBN 978-1-64490-381-0

This book contains information obtained from authentic and highly regarded sources. Reasonable efforts have been made to publish reliable data and information, but the authors and publisher cannot assume responsibility for the validity of all materials or the consequences of their use. The authors and publishers have attempted to trace the copyright holders of all material reproduced in this publication and apologize to copyright holders if permission to publish in this form has not been obtained. If any copyright material has not been acknowledged, please write and let us know so we may rectify in any future reprint.

Distributed worldwide by

Materials Research Forum LLC
105 Springdale Lane
Millersville, PA 17551
USA
https://mrforum.com

Printed in the United States of America
10 9 8 7 6 5 4 3 2 1

Table of Contents

Materials Research Foundations **183** (2025)   https://doi.org/10.21741/9781644903810

CHAPTER 1

The current state of research on the characterization of phosphate layers on the surface of steels for reinforced concrete

P. Lazar[1], A.-M. Cazac[1]*, C. Bejinariu[1,2]

[1]Faculty of Materials Science and Engineering, Gheorghe Asachi Technical University of Iasi, Romania

[2]Academy of Romanian Scientists, Ilfov 3, 050044 Bucharest, Romania

alin-marian.cazac@academic.tuiasi.ro

Abstract

The first chapter presents the current state of research on steels used as reinforcements in construction. The materials used in the reinforcements, their manufacturing process and their specific characteristics according to the standards are presented and characterized. In addition, an analysis of the phosphating process used in the thesis is presented, as well as an overview of the physical, mechanical and chemical properties of steels for concrete reinforcement used in civil engineering.

Keywords

Phosphating, Steels For Reinforced Concrete, Phosphate Layers

1.1 Concrete and reinforcement correlation

1.1.1 Reinforced concrete details

Reinforced concrete is a structural material widely used in civil engineering, such as buildings, dams, bridges, etc. These structures are labour intensive, costly and have a limited lifespan. Reinforced concrete structures can suffer physical damage (cracking, frost, fire), chemical damage (sulphate attack, acid attack, seawater, etc.) and corrosive attack of the reinforcement [1-3].

Steel corrosion is the major threat and damage problem to steel reinforcement in concrete structures worldwide. Apart from the usual corrosion concerns due to general exposure to oxygen and moisture, the corrosion of concrete steel reinforcement can be accelerated by two major factors, namely chloride ion ingress and concrete carbonation [4].

Most studies have analysed ribbed bars. There are a smaller number of applications where, due to the smaller diameter of reinforcing bars, they are used in a smooth form, for example in stirrups. For smooth bars, in addition to corrosion resistance [5, 6], the bar adhesion to the concrete is also investigated, which can be improved or maintained by phosphating [7, 8].

1.1.2 *Corrosion effects on the strength of reinforced concrete*

In 2009, Simescu et al. analysed in detail the corrosion behaviour in an alkaline environment (the environment simulates the interstitial electrolyte of concrete at room temperature and was based on $Ca(OH)_2$ + NaOH + KOH + NaCl) of a zinc phosphate-coated concrete steel. They found that in an alkaline solution, with or without chlorides, the phosphate-coated steel specimen showed better strength than the non-phosphate-coated steel. During the first days of immersion in the alkaline medium (pH = 12.6), there was a slow dissolution of phosphophyllite and a strong dissolution of metallic zinc. The latter formed a hydroxysinate complex in the presence of calcium, followed by the precipitation of calcium as calcium hydroxysinate ($Ca(Zn(OH)_3)_22H_2O$). A dense protective layer was formed. In a solution of chloride ions, at very high concentrations exceeding the chloride threshold tolerated for the initiation of steel corrosion in alkaline environments ($[Cl^-]/[OH^-] > 0.6$), the calcium hydroxysinate film formed by this treatment reduced the aggressiveness of the chlorides and provided effective protection against corrosion of steel reinforcements [6].

Over time, however, the concrete continues to harden and carbonation occurs, making the structure susceptible to corrosion [9-12].

Generalised corrosion, which affects the entire length of the bar uniformly, or pitting corrosion, which affects a specific part of the bar, have important effects on the mechanical behaviour of reinforcing bars [13].

A significant consequence of steel corrosion is the change in the mechanical properties of rebars. Although most studies do not focus on this effect, the rebar corrosion causes changes in the mechanical properties of the material [14-17].

According to the Greek standard [18] and other corresponding European National Standards, the capacity of a steel reinforcement in terms of its mechanical performance is considered unchanged throughout the service life of a reinforced concrete structure. It has been acknowledged [19] that chloride-induced corrosion, defined by the continuous appearance of pitting areas on the steel reinforcement, leads to a significant reduction in the cross-sectional area of the bar [17].

The corrosion effects on the mechanical performance of rebars have been studied in the literature and significant changes in strength and ductility have been found [15-17, 19-27]. Research in this area has also shown the influence of corrosion on the deterioration of steel bars embedded in concrete, reducing the strength of the bond between the bar and the concrete [19, 20, 27-30].

Many local variables, including the mineralogy of the raw material, exposure conditions and traditional construction practices, can influence rebar corrosion. For example, since the early 1970s, sea sand has been widely used for civil engineering in certain cities in the coastal region of Argentina. As a result, there are many structures today that show serious damage caused by rebar corrosion [31, 32].

1.1.3   *Corrosion protection methods for steel used in concrete reinforcement*

Various methods are commonly applied to minimise rebar corrosion, such as cathodic protection [33], realcalcination [34] and coatings on the concrete surface or on the rebar exterior [31, 32, 35, 36]. Another alternative involves inhibitors, which can be cost effective and easy to apply. They can be used in concrete reinforcement by adding the inhibitor with water during concrete preparation or by applying it to the outer surface of the hardened concrete. Reviews of the most common types of inhibitors for concrete repair and the different inhibition mechanisms can be found in the literature [37-39]. The most common mixed inhibitors are based on nitrite ions [40, 41]. However, nitrites must be used with care as they can contaminate the surrounding soil or water.

Other ions have been studied to inhibit steel corrosion, namely chromium, phosphorus, tungsten and molybdenum ions [42-44]. Analysing the studies, it can be observed that phosphates have some interesting advantages, such as low cost and low toxicity.

To evaluate the effectiveness of phosphate ions as inhibitors for reinforced concrete, some authors simulate porous solutions [45-54]. In contrast, only a few articles have reported evaluations of the effectiveness of phosphate ions in mortars or grouts at long exposure times [55, 56]. The interaction of phosphate ions with grouts is complex. Some authors state that phosphates can alter the mechanical properties of concrete or alter curing times as they decompose and precipitate as calcium phosphate, reducing the effectiveness of the inhibitor [47, 57, 58]. However, other authors claim that this inhibitor does not interfere and show that it is effective in mortars [55, 56, 58]. In addition, the optimal phosphate/chloride ratio in mortars is controversial and the mechanism of inhibition is not clear. Some authors propose a dual effect, where calcium phosphate may block pores to prevent diffusion, while phosphate may block cathodic or anodic sites [50, 53-56].

Corrosion of steel reinforcement is a major problem worldwide, as premature failure of concrete structures leads to safety problems and huge repair and replacement costs [59].

In an alkaline environment such as concrete (pH > 10), steel is normally passivated. However, there are two main reasons for the failure of the passivation layer. On the one hand, the ingress of aggressive ions (e.g. chloride ions) can cause localised failure of the passivation layer, leading to localised corrosion. On the other hand, carbonation leads to a decrease in pH - if it falls below 10, a stable passivation layer cannot be maintained - so a generalised corrosion can occur [59].

One method of preventing or limiting corrosion of reinforcing steel is by using corrosion inhibitors. Corrosion inhibitors are defined as chemical compounds that reduce the corrosivity of a given medium when present at relatively low concentrations. Different inhibition mechanisms are known; for example, some inhibitors remove a chemically active substance from the medium by reacting with it (e.g. removal of dissolved oxygen), others form a thin protective layer [60]. Depending on where the inhibitors act, they can be classified as anodic inhibitors, which prevent the oxidation reaction, cathodic inhibitors, which diminish the reduction reaction, or mixed inhibitors, which act on both half-reactions. The effectiveness of the inhibitor therefore depends on the corrosive environment and the metal surface on which it must act [60].

Inhibitors can be used either preventively, mainly as an additive to the mixing water of fresh concrete, or restoratively as a surface-applied inhibitor after the concrete has hardened [61-65].

Common inhibitors are nitrites (calcium nitrite and sodium nitrite) and various organic inhibitors such as amino alcohols and amines (sometimes mixtures are used); sodium monofluorophosphate (MFP) is also regularly used as a surface-applied inhibitor [39, 57, 66-72]. In addition, molybdate, tungstate, chromate and other phosphate anions have been tested for their anti-corrosion properties [42, 44, 73].

Unfortunately, there are some drawbacks to using inhibitors. Firstly, toxicity can limit their use in practical applications. For example, sodium chromate is banned by EU directives (REACH) as it is listed as carcinogenic, mutagenic and toxic to reproduction [74]. Secondly, certain interactions may occur between the inhibitor and concrete constituents, which may influence the properties of the concrete (with positive or negative results) [39, 75] or lead to deactivation of the inhibitor, e.g. when MFP is mixed into fresh concrete, the active component PO_3F^- reacts with calcium ions to form insoluble products [57]. Finally, a certain threshold concentration of inhibitor (in terms of aggressive ion concentration) must always be present - locally - at the steel surface to ensure effective inhibition. If the actual inhibitor concentration is too low, there is a risk that the corrosion process may even be accelerated [39, 57]. This is particularly the case with surface-applied inhibitors, as sufficient inhibitor concentration is achieved at the rebar. It can also be a difficult requirement for mixed inhibitors in the long term, as inhibitors can be washed out or evaporate. For both types of application, this also means that only a (small) percentage of the applied inhibitor is effectively utilised [59].

Phosphating is the process of depositing a layer of soluble phosphate compounds on the surface of a metal by conversion. Although phosphate coatings have been studied since the early nineteenth century, they are not only still being studied, but are an area of interest due to their numerous applications. The advantages of this type of coating are well known, such as the low cost of the deposition process, improved corrosion and wear resistance, and possible adhesion of subsequent coatings such as paint. All of this is leading to studies on how to continuously improve the properties of the phosphate coating by modifying the phosphating process parameters, as well as by modifying/replacing the substances used in the phosphating solutions with "environmentally friendly" solutions. Also because of these advantages, several researchers are investigating the possible use of phosphate coatings in areas such as civil engineering or medicine - biomaterial coatings [76-86].

Zn phosphating is one of the most widely used surface treatments and is generally performed in aqueous acid phosphate solutions containing zinc and phosphoric acid ions and nitrate ions as an accelerator to promote oxidation and dissolution of the metal surface [87, 88]. The phosphate coating is known to develop by nucleation, growth and coalescence of zinc phosphate crystals. The corrosion resistance of the phosphate coating is related to the size and density of the pores in the coating [89]; i.e. the pores provide a pathway for corrosion attack [90].

There are well-known solutions with additives such as Ni^{2+}, Mn^{2+} and Mg^{2+} which contribute to a homogeneous phosphate coating with finer zinc phosphate grains [88, 91-96]. The presence of Ni^{2+} and Mn^{2+} in the phosphate solution appears to affect the nucleation and growth of zinc phosphate crystals.

The crystalline phosphating processes occur through a "common electrode" electrochemical mechanism whereby the anodic processes are of the polyelectrode type (short-circuited micropores) and produce an interaction between the poorly soluble pyrophosphate layer and the iron substrate. The Fe^{2+} ions formed in the anodic region (dissolution of the metal) contribute to the formation of the primary zinc pyrophosphate layer, the resulting crystalline structures becoming inert to subsequent oxidative processes. Hydrogen is released in the cathodic region. The two electrochemical processes may or may not occur simultaneously, resulting in small areas of polarisation with electron transfer through the iron substrate. Generally, the phosphating process occurs in a single step [97-103].

There are many anticorrosive processes for the surface treatment of iron parts by chemical phosphating, which consists in the precipitation of a thin, continuous and uniform layer of low-solubility pyrophosphates of zinc, nickel, manganese or calcium and iron. Zinc is the most commonly used metal. It can be used in a mixture or alone, depending on the coating application [97-103].

The metallic materials are exposed to a continuous process of structural degradation as a result of interactions with the environment, a process known as 'corrosion' [104].

Except for precious metals, all other metals are unstable in contact with atmospheric air. How and to what extent this instability manifests itself depends on both the nature of the metal and its environment. In ferrous materials, for example, iron oxide is formed as a result of corrosion. This corrosion process is called antimetallurgy because it tends to return metals to their natural state [105].

Metal corrosion concerns mainly the energy, transport, chemical, food, petroleum and mechanical sectors [106-108]. The resulting damage is enormous, taking into account the intrinsic value of the corroded metal, its replacement cost and the cost of preventing the destructive process, simply known as direct costs [109, 110].

Indirect costs are those associated with reduced metal life, product loss, pollution, manufacturing failure, sudden damage or explosion [111]. Indirect costs are difficult to predict and often exceed direct costs.

Materials Research Foundations **183** (2025)   https://doi.org/10.21741/9781644903810

The costs associated with metal corrosion are very high. Corrosion leads to high costs due to material degradation, direct corrosion, but especially due to the decommissioning of metal parts and equipment [112, 113]. Corrosion can also cause significant repair costs of equipment components [111].

It is estimated that almost a third of the world's metal production is lost to corrosion. Since only about two-thirds of corroded metal is recovered by melting, this means that about 10% of the world's production is permanently lost to the destructive effects of corrosion [108, 110, 114].

This enormous damage demonstrates the need to systematically investigate the corrosion causes and to develop effective methods to prevent it. The economic impact of corrosion prevention is important. Numerous studies show that the application of corrosion protection leads to large cost savings in shipbuilding, civil engineering, the chemical industry, etc. [115-118].

Therefore, over time, several methods have been developed to protect metals against corrosion, including electrolytic deposition, spray coatings, diffusion coatings, plating coatings, phosphating, oxidisation, enamelling, varnishing, painting, bituminous coatings, physical and chemical vapour deposition methods [78, 79, 83, 119].

From these surface treatments, chemical conversion treatment is a simple and inexpensive method used in a wide range of applications and is one of the most widely used methods of corrosion protection [80, 81]. Phosphating is an anticorrosive process that consists in converting the metal surface into a phosphate film by attacking it with complex solutions containing phosphoric acid and oxides of zinc, manganese, iron, etc. [82].

Being a conversion coating, the phosphate film has a very good adhesion to the supporting metal, ensuring the adhesion of the whole protective system, and by its absorbing and adsorbing properties it improves the quality of the whole protective system [79-83]. Corrosion protection is achieved by the formation of a uniform and compact passivation layer consisting of a mixture of secondary and tertiary phosphates of zinc, manganese, calcium, etc., which are difficult to dissolve, obtained by coprecipitation or sequential precipitation in several stages, after prior pickling of ferrous metal surfaces [79-83].

Applications in the oil and gas industry include corrosion protection [120-123] of casing joints during storage. An additional benefit of coatings is improved wear resistance during assembly [124].

The casing joints contain a metal-to-metal seal to ensure the pressure integrity of the pipe created after installation. The performance of metal-to-metal seals has been shown to be determined by the surface texture and surface modifications [125-128]. Therefore, phosphate coatings play an important role in the sealing ability of casing joints [129].

Priority areas for phosphating process improvement include improving the protective and other functional properties of coatings, reducing solution concentration, temperature and processing time, incorporating customisation, standardising phosphating compositions and reducing the environmental risk of the processes [102].

1.2 Specific phosphating processes

1.2.1 Phosphating solutions. Phosphating parameters

All conventional phosphating solutions are based on dilute phosphoric acid solutions based on alkali or heavy metal ions, containing mainly free phosphoric acid and primary phosphates of the metal ions contained in the bath [65]. Traditionally, phosphating is performed by immersing the products in a phosphate solution bath heated to 80-95°C [102]. The process lasts 40-60 minutes and depends on the concentration of phosphate ions in the working solution and the processing temperature. In order to accelerate the phosphating process, the release of hydrogen and the oxidation of Fe^{2+} to Fe^{3+} are reduced, the oxidants $Zn(NO_3)_2$, NaF, $NaNO_2$ are introduced into the bath [62-65]. Hot solution phosphating consumes a large amount of energy. In addition, the main reaction of hydrolytic decomposition of phosphating preparations occurs even in the absence of coated parts in the bath, which leads to an increase in the free acidity of the solution and, consequently, to an increase in the porosity of the films and a decrease in the protective properties [62-65, 102].

Phosphating at room temperature significantly slows down the reaction rate of hydrolytic decomposition of the phosphating preparations towards the formation of free orthophosphoric acid, which almost eliminates the dissolution of crystalline phosphates formed on the surface during phosphating. The most favorable temperature range for phosphating solutions is 20-30°C [35, 65, 102]. However, the phosphate film formation rate and thickness are significantly reduced in cold solutions. To accelerate the interaction of the steel with the free orthophosphoric acid, oxidizing agents (nitrates, nitrites, fluorides) are added to the phosphating solution. A very important advantage of the cold phosphating process is the control of the process and the possibility of spraying the solution to treat large surfaces and bulky complex structures [51].

The introduction of additional metal salts into phosphating solutions can change the colour of the deposited film [105, 127, 128]. Phosphating compositions containing colourants have been developed for the deposition of blue and green phosphate films [102].

1.2.2 Phosphate coating properties

Phosphate coatings on steel, zinc, galvanised steel, aluminium and other similar metals have a crystalline structure with a crystal size from a few micrometres to about 100 μm [91, 92]. The composition of phosphate coatings includes many different components (Rumyantsev E., 2022). More than 30 phosphate compounds [94] are found in phosphate coatings. The phase components found in the crystalline phosphate film have different colours, which is also reflected in the colour of the film [94].

When phosphating the metal surface, two main processes are observed: phosphate precipitation and dissolution of the base metal. During phosphating, the surface layer of the metal is etched as a result of the interaction with the phosphating solution. The surface of the steel is transformed into a crystalline layer of insoluble secondary and tertiary phosphates which adhere to and are an integral part of the base metal. The main advantages of phosphate coatings are good adhesion to the metal surface and high protective properties [7].

1.2.3   Applications

Three types of phosphate coatings are widely used: based on iron, manganese and zinc phosphates [93].

Iron phosphate coatings were the first to be developed. Such phosphate coatings are dark grey in colour. They have a wide range of applications for surface treatment of steel products prior to painting [111, 112].

Manganese phosphate coatings are dark grey and black in colour. Manganese phosphate coatings are used to reduce product friction in combination with lubrication [2, 97, 101].

The colour of zinc phosphate films can vary from light to dark grey. The difference in the colour is caused not only by the chemical composition of the phosphating solution, but also by the chemical composition of the steel and the preparation of the metal surface prior to the precipitation of the phosphate film. This type of phosphate coating is applied to the surface of steel products not only for corrosion protection, but also to increase wear resistance, improve machinability and prior to painting or oiling [127-130].

Phosphate coatings are usually obtained by immersing a steel product in a solution, by spraying a phosphating solution onto the surface of the product, or by applying the solution to the surface with a pad or sponge [102]. The selection of the method depends on the size of the product and the complexity of the product profile [102].

1.3   Conclusions

From the literature review, the following conclusions can be drawn:

1. To improve the corrosion resistance of steels intended for concrete reinforcement, a series of studies on the production of phosphate coatings is required. This requires the optimisation of phosphating solutions and the development of phosphate coatings using a specific technology.

2. After obtaining phosphate layers on concrete steels, it is necessary to study the morphology and structure of the deposited phosphate layers, as follows:

- structural characterisation of the phosphate layers by scanning electron microscopy;

- determination of the chemical composition of the phosphate layers by EDX analysis;

- determination of the chemical compounds formed after phosphating by X-ray diffraction.

3. The characterisation of the mechanical properties of the deposited layers is essential:

- determination of mechanical properties by roughness tests;

- determination of mechanical properties by scratch testing;

- determination of mechanical properties by micro-indentation testing.

4. The corrosion behaviour of the deposited layer is an essential parameter. It significantly influences the quality of the concrete obtained;

5. It is also essential to determine the bond between the reinforcement (reinforcement with a phosphate coating) and the concrete.

References

[1]   Gani, M.S.J. Fiber Reinforced Cement and Concrete Composites; Chapman & Hall: Sarasota, FL, USA, 1997; pp. 128–145.

[2]   Mehta, K.; Monteiro, P. Concrete: Microstructure, Properties and Materials; McGraw-Hill Education: Chicago, IL, USA, 2014; ISBN: 9780071797870.

[3]   Duffó, G.; Reinoso, M.; Ramos, C.; Farina, S. Characterization of steel rebars embedded in a 70-year old concrete structure. Cem. Concr. Res. 2012, 42, 111–117. https://doi.org/10.1016/j.cemconres.2011.08.003

[4]   Shin, C.B.; Kim, E.K. Modeling of chloride ion ingress in coastal concrete. Cem. Concr. Res. 2002, 32, 757–762. https://doi.org/10.1016/S0008-8846(01)00756-6

[5]   Yohai, L.; Valcarce, M.; Vázquez, M. Testing phosphate ions as corrosion inhibitors for construction steel in mortars. Electrochim. Acta 2016, 202, 316–324. https://doi.org/10.1016/j.electacta.2015.12.124

[6]   Simescu, F.; Idrissi, H. Corrosion behaviour in alkaline medium of zinc phosphate coated steel obtained by cathodic electro-chemical treatment. Corros. Sci. 2009, 51, 833–840. https://doi.org/10.1016/j.corsci.2009.01.010

[7]   Samardžija M.; Alar V.; Aljinović F.; Kapor F., Influence of phosphate layer on adhesion properties between steel surface and organic coating. Rud. Geološko Naft. Zb. 2022, 37, 11–17. https://doi.org/10.17794/rgn.2022.1.2.55

[8]   Bajat, J.; Mišković-Stanković, V.; Popić, J.; Dražić, D. Adhesion characteristics and corrosion stability of epoxy coatings electro-deposited on phosphated hot-dip galvanized steel. Prog. Org. Coat. 2008, 63, 201–208. https://doi.org/10.1016/j.porgcoat.2008.06.002

[9]   Marcos-Meson, V.; Michel, A.; Solgaard, A.; Fischer, G.; Edvardsen, C.; Skovhus,

T.L. Corrosion resistance of steel fibre rein-forced concrete—A literature review. Cem. Concr. Res. 2018, 103, 1–20.

[10]   Lazar, P.; Bejinariu, C.; Cazac, A.M.; Sandu, A.V.; A Bernevig, M.; Burduhos-Nergis, D.P. Phosphate coatings for the protection of steels reinforcement for concrete, 2013, J. Phys. 1960, 1960, 012013. https://doi.org/10.1088/1742-6596/1960/1/012013

[11]   Jiang, C.; Gao, Z.; Pan, H.; Cheng, X. The initiation and formation of a double-layer phosphate conversion coating on steel. Electrochem. Commun. 2020, 114, 106676. https://doi.org/10.1016/j.elecom.2020.106676

[12]   Ezekiel, S.; Ayoola, A.; Durodola, B.; Odunlami, O.; Olawepo, A. Data on zinc phosphating of mild steel and its behaviour. CDC 2022, 38, 100838. https://doi.org/10.1016/j.cdc.2022.100838

[13]   Fernandez, I.; Bairán, J.M.; Marí, A.R. Corrosion effects on the mechanical properties of reinforcing steel bars. Fatigue and σ-ε behavior. Constr. Build Mater. 2015, 101, 772–783. https://doi.org/10.1016/j.conbuildmat.2015.10.139

[14]   Burduhos-Nergis D.P., Cazac A.M., Corabieru A., Matcovschi E., Bejinariu C., Characterization of Zinc and Manganese Phosphate Layers Deposited on the Carbon Steel Surface, 2020. IOP Conference Series: Materials Science and Engineering 877, 012012.

[15]   American Standards Association, American Society of Mechanical Engineers, Society of Automotive Engineers, Surface Texture: Surface Roughness, Waviness, and Lay; American Society of Mechanical Engineers: New York, NY, USA, 2020; ISBN 978-0-7918-7325-0. OCLC 1197629204

[16]   Zhang, W.; Song, X.; Gu, X.; Li, S. Tensile and fatigue behavior of corroded rebars. Constr. Build. Mater. 2012, 34, 409–417. https://doi.org/10.1016/j.conbuildmat.2012.02.071

[17]   Apostolopoulos, C.A., Mechanical behavior of corroded reinforcing steel bars S500s tempcore under low cycle fatigue. Constr. Build. Mater. 2007, 21, 1447–1456. http://doi.org/10.1016/j.conbuildmat.2006.07.008

[18]   Apostolopoulos, C.; Papadopoulos, M.; Pantelakis, S. Tensile behaviour of corroded reinforcing steel bars BSt 500s. Constr. Build. Mater. 2006, 20, 782–789.

[19]   Apostolopoulos, C.; Papadakis, V. Consequences of steel corrosion on the ductility properties of reinforcement bar. Constr. Build. Mater. 2008, 22, 2316–2324.

[20]   Apostolopoulos, C.A.; Demis, S.; Papadakis, V.G. Chloride-induced corrosion of steel reinforcement—Mechanical performance and pit depth analysis. Constr. Build. Mater. 2013, 38, 139–146.

[21]   Demis, S.; Pilakoutas, K.; Apostolopoulos, C.A. Effect of corrosion on bond strength of steel and non-metallic reinforcement. Mater. Corros. 2010, 61, 328–

331.

[22]   Batis G., Rakanta E., Corrosion of steel reinforcement due to atmospheric
        pollution. Cem. Concr. Compos. 2005, 27, 269–275.

[23]   Moreno, E.; Cobo, A.; Cánovas, M.F. Mechanical properties variation of B500SD
        high ductility reinforcement regarding its corrosion degree. Mater. Constr. 2011,
        61, 517–532.

[24]   Fang, C.; Lundgren, K.; Chen, L.; Zhu, C. Corrosion influence on bond in
        reinforced concrete. Cem. Concr. Res. 2004, 34, 2159–2167.

[25]   Apostolopoulos, C.A. The influence of corrosion and cross-section diameter on the
        mechanical properties of B500c steel. J. Mater. Eng. Perform. 2009, 18, 190–195.

[26]   Cairns, J.; Plizzari, G.A.; Du, Y.; Law, D.W.; Frnazoni, C. Mechanical properties
        of corrosion-damaged reinforcement. ACI Mater. J. 2005, 102, 256–264.

[27]   Almusallam, A.A. Effect of degree of corrosion on the properties of reinforcing
        steel bars. Constr. Build. Mater. 2001, 15, 361–368.

[28]   Capozucca, R. Damage to reinforcement concrete due to reinforcement corrosion.
        Constr. Build. Mater. 1995, 9, 295–303.

[29]   Du, Y.G.; Clark, L.A.; Chan, A.H.C. Effect of corrosion on ductility of reinforcing
        bars. Mag. Concr. Res. 2005, 57, 407–419.

[30]   Du, Y.G.; Clark, L.A.; Chan, A.H.C. Residual capacity of corroded reinforcing
        bars. Mag. Concr. Res. 2005, 57, 135–147.

[31]   Morris W., Vázquez M., Corrosion of reinforced concrete exposed to marine
        environment, 2002, Corrosion Reviews 20, 469–508,
        https://doi.org/10.1515/CORRREV.2002.20.6.469

[32]   Morris W., Vázquez M., Sánchez S.R.d., Efficiency of coatings applied on rebars
        in concrete, 2000, Jounal of Materials Science 35, 1885–1890,
        https://doi.org/10.1023/A:1004745831261

[33]   Page C.L., Ngala V.T., Page M.M., Corrosion inhibitors in concrete repair
        systems, 2000, Magazine of Concrete Research 52, 25,
        https://doi.org/10.1680/macr.2000.52.1.25

[34]   Page C.L., Sergi G., Developments in cathodic protection applied to reinforced
        concrete, 2000, Journal of Materials in Civil Engineering 12, 8–15,
        https://doi.org/10.1061/(ASCE)0899-1561(2000)12:1(8)

[35]   Sagües A.A., Perez Duran H.M., Powers R.G., Corrosion Performance of Epoxy-
        Coated Reinforcing Steel in Marine Substructure Service, 1991, Corrosion 47,
        884–893, https://doi.org/10.5006/1.3585202

[36]   Jalili M.M., Moradian S., Hosseinpour D., The use of inorganic conversion
        coatings to enhance the corrosion resistance of reinforcement and the bond

strength at the rebar/concrete, 2009, Construction and Building Materials 23, 233–238, https://doi.org/10.1016/j.conbuildmat.2007.12.011

[37]   Gaidis J.M., Chemistry of corrosion inhibitors, Cement & Concrete Composites, 2004 26, 181, https://doi.org/10.1016/S0958-9465(03)00037-4

[38]   Sastri V.S., Corrosion Inhibitors. Principles and Applications, 1998, John Wiley & Sons, West Sussex, England, ISBN 10: 0471976083, ISBN 13: 9780471976080.

[39]   Söylev T., Richardson M., Corrosion inhibitors for steel in concrete: state of the art report, Constr. Build. Mater. 22 (4) (2008) 609–622, https://doi.org/10.1016/j.conbuildmat.2006.10.013

[40]   Valcarce M.B., Vazquez M., Carbon steel passivity examined in alkaline solutions: the effect of chloride and nitrite ions, Electrochimica Acta, Volume 53, Issue 15, 2008, Pages 5007-5015, https://doi.org/10.1016/j.electacta.2008.01.091

[41]   Hansson C.M., Mammoliti L., Hope B.B., Corrosion inhibitors in concrete. Part I: the principles, Cement & Concrete Composites Volume 28, Issue 12, December 1998, Pages 1775-1781, https://doi.org/10.1016/S0008-8846(98)00142-2

[42]   Abd El Haleem S.M., Abd El Wanees S., Abd El Aal E.E., Diab A., Environmental factors affecting the corrosion behavior of reinforcing steel II. Role of some anions in the initiation and inhibition of pitting corrosion of steel in Ca(OH)2 solutions, Corros. Sci. 52 (5) (2010) 292–302, https://doi.org/10.1016/j.corsci.2010.01.021

[43]   Gao Y.B., Hu J., Zuo J., Liu Q., Zhang H., Dong S.G, Du R.G, Lin C.J., Synergistic inhibition effect of sodium tungstate and hexamethylene tetramine on reinforcing steel corrosion, Journal of the Electrochemical Society 162 (2015) C555–C562, https://doi.org/10.1149/2.0641510jes

[44]   Tang Y., Zhang G., Zuo Y., The inhibition effects of several inhibitors on rebar in acidified concrete pore solution, Constr. Build. Mater. 28 (1) (2012) 327–332, https://doi.org/10.1016/j.conbuildmat.2011.08.048

[45]   Etteyeb N., Dhouibi L., Takenouti H., Alonso M.C., Triki E., Corrosion inhibition of carbon steel in alkaline chloride media by Na3PO4, Electrochimica Acta 52 (2007) 7506–7512.

[46]   Etteyeb N., Sanchez M., Dhouibi L., Alonso C., Andrade C., Triki E., Corrosion protection of steel reinforcement by a pretreatment in phosphate solutions: Assessment of passivity by electrochemical techniques, Corrosion Engineering Science and Technology 41 (2006) 336–341.

[47]   Etteyeb N., Sánchez M., Dhouibi L., Alonso M.C., Takenouti H., Triki E., Effectiveness of pretreatment method to hinder rebar corrosion in concrete, Corrosion Engineering, Science and Technology 45 (2010) 435–441.

[48]   Génin J.M.R., Dhouibi L., Refait P., Abdelmoula M., Triki E., Influence of

phosphate on corrosion products of iron in chloride-polluted-concrete- simulating solutions: Ferrihydrite vs green rust, Corrosion 58 (2002) 467–478.

[49]   Bastidas D.M., Criado M., Fajardo S., La Iglesia A., Bastidas J.M., Corrosion inhibition mechanism of phosphates for early-age reinforced mortar in the presence of chlorides, Cement and Concrete Composites 61 (2015) 1–6.

[50]   Bastidas D.M., Criado M., La Iglesia V.M., Fajardo S., La Iglesia A., Bastidas J.M., Comparative study of three sodium phosphates as corrosion inhibitors for steel reinforcements, Cement & Concrete Composites 43 (2013) 31–38.

[51]   Reffass M., Sabot R., Jeannin M., Berziou C., Refait P., Effects of phosphate species on localised corrosion of steel in NaHCO3 + NaCl electrolytes, Electrochimica Acta 54 (2009) 4389–4396.

[52]   Yohai L., Vázquez M., Valcarce M.B., Phosphate ions as corrosion inhibitors for reinforcement steel inchloride-rich environments, Electrochimica Acta 102 (2013) 88–96.

[53]   Nahali H., Dhouibi L., Idrissi H., Effect of Na3PO4 addition in mortar on stee reinforcement corrosion behavior in 3% NaCl solution, Construction and Building Materials. 78 (2015) 92–101.

[54]   Nahali H., Dhouibi L., Idrissi H., Effect of phosphate based inhibitor on the threshold chloride to initiate steel corrosion in saturated hydroxide solution, Construction and Building Materials 50 (2014) 87–94.

[55]   Dhouibi L., Triki E., Salta M., Rodrigues P., Raharinaivo A., Studies on corrosion inhibition of steel reinforcement by phosphate and nitrite, Materials and Structures/Materiaux et Constructions 36 (2003) 530–540.

[56]   Shi J.J., Sun W., Effects of phosphate on the chloride-induced corrosion behavior of reinforcing steel in mortars, Cement & Concrete Composites 45 (2014) 166–175.

[57]   Elsener B., Corrosion inhibitors for steel in concrete, in: Materials Week, International Congress on Advanced Materials, 25-28.9.2000 ICM International Congress Centre Munich. Session E1 Corrosion of Steel in Concrete, 2000.

[58]   Elsener B., Corrosion inhibitors for steel in concrete. State of the art report, Maney Publishing, London, 2001.

[59]   Verbruggen H., Terryn H., De Graeve I., Inhibitor evaluation in different simulated concrete pore solution for the, protection of steel rebars, Construction and Building Materials 124 (2016) 887–896.

[60]   Callister Jr. W.D., Materials Science and Engineering – An Introduction, seventh ed., John Wiley & Sons, Inc., 2007, ISBN-13: 978-0-471-73696-7.

[61]   Lazar, P., Bejinariu, C., Cazac, A.M., Sandu, A.V., Bernevig, M.A., Burduhos-Nergis, D.P. Phosphate coatings for the protection of steels reinforcement for

concrete. (2021) Journal of Physics: Conference Series, 1960 (1), art. no. 012013. DOI: 10.1088/1742-6596/1960/1/012013.

[62]　Lazar P, Bejinariu C, Sandu AV, Cazac AM, Sandu IG, Chemical Deposition of Thin Layers on Reinforcing Steel. International Conference on Innovative Research, May 14th to 15th, ICIR 2015, Iasi – Romania, Key Engineering Materials Vol 660 (2015) pp 213-218, (2015) Trans Tech Publications, Switzerland, https://doi.org/10.4028/www.scientific.net/KEM.660.213

[63]　Lazar P., Cimpoesu N., Istrate B., Cazac A.M., Burduhos-Nergis D.-P., Benchea M., Berbecaru A.C., Badarau G., Vasilescu G.D., Popa M. and Bejinariu C. Microstructural and Mechanical Properties Analysis of Phosphate Layers Deposited on Steel Rebars for Civil Constructions. Coatings (Coatings), 2024, Volume 14, Issue 2, Article Number 182. DOI https://doi.org/10.3390/coatings14020182

[64]　Lazar, P., Bejinariu, C., Sandu, A.V., Cazac, A.M., Corbu, O., Sandu, I.G., Perju, M.C. Corrosion Evaluation of Some Phosphated Thin Layers on Reinforcing Steel. (2017) IOP Conference Series: Materials Science and Engineering, 209 (1), art. no. 012025. https://doi.org/10.1088/1757-899X/209/1/012025, eISSN 1757-899X

[65]　Lazar, P., Cazac, A.M., Perju, M.C., Bujoreanu, L.G., Bejinariu, C. Phosphate coatings suitable for steel rebars used for reinforced concrete. Buletinul Institutului Politehnic din Iaşi publicat de Editura POLITEHNIUM din Iaşi, Volumul 69 (73), Numărul 1-4, Secţia Ştiinţa şi Ingineria Materialelor, 2023, pp 185-193, ISSN 1453-1690.

[66]　Bertolini L., Elsener B., Pedeferri P., Redaelli E., Polder R.B., Corrosion of Steel in Concrete: Prevention, Diagnosis, Repair, John Wiley & Sons Inc, 2013.

[67]　Ramasubramanian B.M., Haran B.S., Popova S., Popov B.N., Petrou M.F., Member A., White R.E., Inhibiting action of calcium nitrite on carbon steel rebars, J. Mater. Civ. Eng. (February) (2001) 10–17.

[68]　Valcarce M.B., Lopez C., Vazquez M., The role of chloride, nitrite and carbonate ions on carbon steel passivity studied in simulating concrete pore solutions, J. Electrochem. Soc. 159 (5) (2012) C244, https://doi.org/10.1149/2.006206jes

[69]　Królikowski A., Kuziak J., Impedance study on calcium nitrite as a penetrating corrosion inhibitor for steel in concrete, Electrochim. Acta 56 (23) (2011) 7845–7853, https://doi.org/10.1016/j.electacta.2011.01.069

[70]　Dong B., Wang Y., Ding W., Li S., Han N., Xing F., Lu Y., Electrochemical impedance study on steel corrosion in the simulated concrete system with a novel self-healing microcapsule, Constr. Build. Mater. 56 (2014) 1–6, https://doi.org/10.1016/j.conbuildmat.2014.01.070

[71]　Fei F.-L., Hu J., Wei J.-X., Yu Q.-J., Chen Z.-S., Corrosion performance of steel reinforcement in simulated concrete pore solutions in the presence of imidazoline

quaternary ammonium salt corrosion inhibitor, Constr. Build. Mater. 70 (2014) 43–53, https://doi.org/10.1016/j.conbuildmat.2014.07.082

[72]   Sun Q.L., Liu Z.R., Guo Y.Z., Liu F., Effect of new corrosion inhibitor on electrochemical behavior of prestressed steel wire in simulated concrete pore solution, Adv. Mater. Res. 941–944 (2014) 1390–1393, https://doi.org/10.4028/www.scientific.net/AMR.941-944.1390

[73]   Tommaselli M., Mariano N., Kuri S., Effectiveness of corrosion inhibitors in saturated calcium hydroxides solutions acidified by acid rain components, Constr. Build. Mater. 23 (1) (2009) 328–333, https://doi.org/10.1016/j.conbuildmat.2007.12.002

[74]   ECHA (European Chemicals Agency), Member state committee draft support document for identification of boric acid as a substance of very high concern because of its CMR properties, SVHC Support Document (June) (2010) pp. 1–27.

[75]   Fayala I., Dhouibi L., Nóvoa X.R., Ben Ouezdou M., Effect of inhibitors on the corrosion of galvanized steel and on mortar properties, Cem. Concr. Compos. 35 (1) (2013) 181–189, https://doi.org/10.1016/j.cemconcomp.2012.08.014

[76]   Burduhos Nergis D.P., Burduhos Nergis D.D., Baltatu M.S., Vizureanu P., Advanced Coatings for the Corrosion Protection of Metals, 2022, Materials Research Forum 115.

[77]   Burduhos Nergis D.P., Cimpoesu N., Vizureanu P., Baciu C., Bejinariu C., Tribological characterization of phosphate conversion coating and rubber paint coating deposited on carbon steel carabiners surfaces, 2019, Materials Today: Proceedings 19, 969-78.

[78]   Burduhos Nergis D.P., Nejneru C., Burduhos Nergis D.D., Savin C., Sandu A.V., Toma S.L. Bejinariu C., The galvanic corrosion behavior of phosphated carbon steel used at carabiners manufacturing, 2019, Revista de Chimie 70, 215-9.

[79]   Burduhos-Nergis D.P., Bejinariu C., Sandu A.V., Phosphate Coatings Suitable for Personal Protective Equipment, 2021, Materials Research Forum 89.

[80]   Burduhos-Nergis D.P., Bejinariu C., Toma S.L., Tugui A.C., Baciu E.R., Carbon steel carabiners improvements for use in potentially explosive atmospheres, 2020, MATEC Web of Conferences 305, 00015.

[81]   Burduhos-Nergis D.-P., Sandu A.V., Burduhos-Nergis D.-D., Vizureanu P., Bejinariu C., Phosphate Conversion Coating – A Short Review, Arch. Metall. Mater. 68 (2023), 3, 1029-1034, https://doi.org/10.24425/amm.2023.145471

[82]   Burduhos-Nergis D.P., Vasilescu G.D., Burduhos-Nergis D.D., Cimpoesu R., Bejinariu C., Phosphate coatings: EIS and SEM applied to evaluate the corrosion behavior of steel in fire extinguishing solution, 2021, Applied Sciences 11 (17), 7802.

[83] Burduhos-Nergis D.P., Vizureanu P., Sandu A.V., Bejinariu C., Phosphate Surface Treatment for Improving the Corrosion Resistance of the C45 Carbon Steel Used in Carabiners Manufacturing, 2020, Materials 13, 3410.

[84] Burduhos-Nergis, D.-P.; Vizureanu, P.; Sandu, A.V.; Bejinariu, C. Phosphate surface treatment for improving the corrosion resistance of the c45 carbon steel used in carabiners manufacturing. Materials 2020, 13, 3410. https://doi.org/10.3390/ma13153410

[85] Bejinariu C, Lazar P, Sandu AV, Cazac AM, Sandu IG, Corbu O, Enhancing Properties of Reinforcing Steel by Chemical Phosphatation. The ICAMET 2014, 3th International Conference Proceedings – Advanced Materials Engineering & Technology, Ho Chi Minh City, Vietnam, December 4-5, 2014, Applied Mechanics and Materials Vols. 754-755 (2015) pp 310-314, (2015) Trans Tech Publications, Switzerland, https://doi.org/10.4028/www.scientific.net/AMM.754-755.310, ISBN: 978-3-03835-434-5

[86] Bejinariu C., Burduhos-Nergis D.P., Cimpoesu N., Immersion Behavior of Carbon Steel, Phosphate Carbon Steel and Phosphate and Painted Carbon Steel in Saltwater, 2021, Materials 14, 188.

[87] Zimmermann D., Munoz A.G., Schultze J.W., Formation of Zn–Ni alloys in the phosphating of Zn layers, Surface and Coatings Technology, 197, 2005, 260–269.

[88] Zimmermann D., Munoz A.G., Schultze J.W., Microscopic local elements in the phosphating process, Electrochimica Acta, 48, 2003, 3267–3277.

[89] Banczek E.P., Rodrigues P.R.P., Costa I., Evaluation of porosity and discontinuities in zinc phosphate coating by means of voltametric anodic dissolution (VAD), Surface and Coatings Technology, 203, 2009, 1213–1219.

[90] Creus J., Mazille H., Idrissi H., Porosity evaluation of protective coatings onto steel, through electrochemical techniques, Surface and Coatings Technology, 130, 2000, 224–232

[91] Satoh N., Effects of heavy metal additions and crystal modification on the zinc phosphating of electrogalvanized steel sheet, Surface and Coatings Technology, 30, 1987, 171–181.

[92] Satoh N., Minami T., Relationship between the formation of zinc phosphate crystals and their electrochemical properties, Surface and Coatings Technology, 34, 1988, 331–343.

[93] Ogle K., Tomandl A., Meddahi N., Wolpers M., The alkaline stability of phosphate coatings I: ICP atomic emission spectroelectrochemistry, Corrosion Science, 46, 2004, 979–995.

[94] Sato N., Minami T., Kono H., State analysis related to zinc and manganese components in hopeite crystals of zinc phosphate films by XANES and EXAFS,

Surface and Coatings Technology, 37 (1989) 23–30.

[95] Liu J., Qu J., Zhou X., Liu P., Yang Y., Patent CN101684389 (A),2010.

[96] Ishizuka K., Shindo H., Hayashi K., US Patent 6596,414 B1,2003.

[97] Marinescu A., Andonianţ Gh., Bay, E., Electrochemical and Chemical Technologies of Protection of Metallic Materials, Bucureşti, Ed. Tehnică, 1984.

[98] Townsend H.E., NACE Annual Corrosion Conference, Cincinnati, OH, 1991. Paper No. 416.

[99] Oniciu L., Grüwald E., Galvanotechnica, Bucuresti, Ed. Ştiinţifică şi Enciclopedică, 1980.

[100] Amirudin A., Thierry D., Corrosion mechanisms of phosphated zinc layers on steel as substrates for automotive coatings, Progress in Organic Coatings, 28, 1996, 59-76.

[101] Martin J.M., Antiwear mechanisms of zinc dithiophosphate: a chemical hardness approach, Tribology Letters, 6, 1, 1999, 1-8.

[102] Rausch W., The Phosphating of Metals, Finishing Publications Ltd. (UK), 1990.

[103] Ghali E.I., Potvin R.J.A., The mechanism of phosphating of steel, Corrosion Science, 12, 7, 1972, 583-594.

[104] Michailidis N., Castaneda H., Corrosion, CIRP Encyclopedia of Production Engineering (2018)1-8

[105] Philip A., Schweitzer P.E., Metallic Materials: Physical, Mechanical and Corrosion Properties, New York: Marcel Dekker INC. (2003), ISBN: 0-8247-0878-4.

[106] Valdez B., Schorr M., Zlatev R., Carrillo M., Stoytcheva M., Alvarez L., Eliezer A., Rosas N., Environmental and Industrial Corrosion – Practical and Theoretical Aspects, 2012, Corrosion Control in Industry.

[107] Anon, Research Opportunities in Corrosion Science and Engineering, Study Report 1-176 (2011).

[108] Anon, The Effects and Economic Impact of Corrosion, 2000, Report.

[109] Baker M., R. Fessler, Pipeline Corrosion,2008, Final Report.

[110] Bowman E., J. Varney, N. Thompson, O. Moghissi, M. Gould, J. Payer, International Measures of Prevention, 2016, Application, and Economics of Corrosion Technologies Study, Houston, Texas, USA.

[111] Norgate T.E., Jahanshahi S., Rankin W.J., Assessing the environmental impact of metal production processes, 2007, Journal of Cleaner Production 15, 838-48.

[112] Novák P., Environmental deterioration of metals, 2007, WIT Press 28.

[113] Popoola L.T., Grema A.S., Latinwo G.K., Gutti B., Balogun A.S., Corrosion problems during oil and gas production and its mitigation, 2013, International Journal of Industrial Chemistry 4, 1-15.

[114] Harsimran S., Santosh K., Rakesh K., Overview of corrosion and its control: a critical review, Proceedings on Engineering Sciences, 2021, 3, 13-24.

[115] Bennett L.H., Kruger J., Parker R.L., Passaglia E., Reimann C., Ruff A.W., Yakowitz H., Economic effects of metallic corrosion in the United States Part I, 1978, NBS Special Publication 511-1.

[116] Zhang F., Pan J., Recent Development of Corrosion Protection Strategy Based on Mussel Adhesive Protein, Frontiers in Materials, 2019, 6, 207.

[117] Bhaskaran R., Palaniswamy N., Rengaswamy N.S., Jayachandran M., A review of differing approaches used to estimate the cost of corrosion (and their relevance in the development of modern corrosion prevention and control strategies), 2005, Anti-Corrosion Methods and Materials 52, 29-41.

[118] Hou B., Li X., Ma X., Du C., Zhang D., Zheng M., Xu W., Lu D., Ma F., The cost of corrosion in China, Materials Degradation 1, 2017, 1-10.

[119] Arthanareeswari M., Kamaraj P., Tamilselvi M., Anticorrosive performance of zinc phosphate coatings on mild steel developed using galvanic coupling, Journal of Chemistry 673961 2013.

[120] Burke D., The sliding friction of bonded solid lubricants [Ph.D. thesis], University of central Lancashire; 2005, URL http://clok.uclan.ac.uk/8690/1/David BurkeMay 05.

[121] Khaleghi M., Gabe D., Richardson M., Characteristics of manganese phosphate coatings for wear-resistance applications. Wear 1979;55(2):277–87. https://doi.org/10.1016/0043-1648(79)90159-5

[122] Totik Y., The corrosion behaviour of manganese phosphate coatings applied to AISI 4140 steel subjected to different heat treatments, Surf Coat Technol 2006;200(8).

[123] Weng D., Jokiel P., Uebleis A., Boehni H., Corrosion and protection characteristics of zinc and manganese phosphate coatings. Surf Coat Technol 1997;88(1–3):147–56.

[124] Ertas A., Experimental investigation of galling resistance in OCTG connections. J Manuf Sci Eng 1992; 114(1):100. https://doi.org/10.1115/1.2899745.udin

[125] Perez-Rafols F., Larsson R., Almqvist A., Modelling of leakage on metal-to-metal seals, Tribol Int 2016;94:421–7. https://doi.org/10.1016/j.triboint.2015.10.003

[126] Perez-Rafols F., Larsson R., Lundstrom S., Wall P., Almqvist A., A stochastic two-scale model for pressure-driven flow between rough surfaces. Proc R Soc A: Math Phys Eng Sci 2016;472(2190):20160069.

https://doi.org/10.1098/rspa.2016.0069.

[127] Perez-Rafols F., Larsson R., van Riet E.J., Almqvist A., On the flow through plastically deformed surfaces under unloading: a spectral approach. Proc Inst Mech Eng Part C: J Mech Eng Sci 2017 (095440621769018)

[128] Inose K., Sugino M., Goto K., Influence of grease on high-pressure gas tightness by metal-to-metal seals of premium threaded connections. Tribol Online 2016;11(2):227–34. https://doi.org/10.2474/trol.11.227

[129] Ernens D., de Rooij M.B., Pasaribu H.R., van Riet E.J., van Haaften W.M., Schipper D.J., Mechanical characterization and single asperity scratch behaviour of dry zinc and manganese phosphate coatings, Tribology International 118 (2018) 474–483, https://doi.org/10.1016/j.triboint.2017.04.034

[130] Seo, Y.H. A study on improving the prediction accuracy of cold forging die life based on quantitative evaluation of phosphate film damage. Sci. Rep. 2023, 13, 16464

CHAPTER 2

Objectives and experimental research methodology

P. Lazar[1], A.-M. Cazac[1]*, C. Bejinariu[1,2]

[1]Faculty of Materials Science and Engineering, Gheorghe Asachi Technical University of Iasi, Romania

[2]Academy of Romanian Scientists, Ilfov 3, 050044 Bucharest, Romania

alin-marian.cazac@academic.tuiasi.ro

Abstract

The second chapter presents the methodology used in the experimental research, the objectives proposed in the thesis, the choice of material, the method of layer deposition, the type and composition of the phosphating solutions, as well as the type of analysis used to highlight the success of the deposited layer and the new properties obtained by the phosphating process. Finally, the equipment used to perform the experiments and to evaluate the results is presented.

Keywords

Phosphating, Experimental, Phosphate Layers, Metodology

2.1　Objectives

The central objective is to enhance the corrosion resistance properties of the OB37 steel, which is intended for use in reinforced concrete, and to devise a method to protect it against corrosion. In order to achieve this objective, a series of steps must be completed:

(i) A thorough analysis of the base material in terms of its chemical composition and structural characterization is imperative;

(ii) preparing the samples according to the planned tests and preparing solutions for the phosphating process;

(iii) phosphating the samples;

(iv) the characterisation of the layers obtained by conducting various laboratory tests, the purpose of which is to ascertain the properties obtained.

The following laboratory investigations were conducted:

- elemental analysis was required to ascertain the chemical composition of the base material, as well as the composition of the subsequently deposited layers.

- structural characterization - was performed to study the microstructure of the metal, as well as the formation of crystals by phosphating.

- mechanical characterization was employed to determine the coefficient of friction and modulus of elasticity of the phosphate layers, as well as to quantify the energy absorbed upon impact;

- chemical characterisation (corrosion resistance) was employed on both the base material and layers deposited in several solutions that may come into contact with the base material;

- investigation of the adhesion between the concrete and the phosphate reinforcement.

2.2 Experimental research methodology

As indicated by the results of this study, the focus on this topic within the context of a PhD thesis has been demonstrated to facilitate enhancements in the properties of OB37 steel. This, in turn, has been shown to result in an improvement in the corrosion behaviour of the steel, thereby mitigating the disadvantages associated with its low corrosion resistance and impact strength properties.

The experimental tests on phosphatized coatings that were conducted within the scope of the present thesis are:

- structural characterization:

- optical microscopy;
- scanning electron microscopy (SEM);
- X-ray diffraction (XRD);

- mechanical characterization:

- scratch test;
- indentation test;

- corrosion resistance:

- linear polarisation;
- cyclic polarisation;

- surface characterization:

- concrete-rebar adhesion.

2.2.1 Material used for the experimental research

For the experimental tests, 8 mm and 10 mm diameter OB37 bars were procured from a specialised distributor and utilised as a substrate for various coatings.

The OB37 concrete steel type is a carbon steel with a smooth profile that is hot rolled, and is most often used to make stirrups that support reinforcements.

The applications of this steel encompass concrete reinforcement, elements and structures composed of reinforced and compressed concrete, civil and industrial construction, and stirrup manufacturing for reinforced concrete structures. In comparison with corrugated sheet, this concrete reinforcing steel exhibits superior malleability.

The utilisation of the materials employed in manufacturing reinforced concrete is contingent upon their possession of favourable adhesion and corrosion resistance properties, in consideration of the stipulated operational conditions.

The material designated as OB37, intended for use in reinforced concrete, consists of bars measuring 12 m in length and with diameters ranging from 8-32 mm, as stipulated in SR 438-1:2012.

For the experimental investigations, OB37 steel bars with a diameter of 10 mm – as illustrated in Fig. 2.1 – have been employed.

Figure 2.1. The samples before phosphating.

The mechanical characteristics of the OB37 steel utilised in the experimental investigations are presented in Table 2.1.

Table 2.1.   Mechanical characteristics of OB37 steel.

OB 37	R_c $[\frac{N}{mm^2}]$	R_m $[\frac{N}{mm^2}]$	Elongation A_5 [%]
	255 (d=6-12)	360	25
	235 (d=14-40)	360	25

The chemical composition of steel, as determined by spark spectroscopy and provided by the supplier, is presented in Table 2.2.

Table 2.2.   Chemical composition of OB37 steel.

OB 37	C	Si	Mn	S	P	Fe
	0.23	0.07	0.75	0.045	0.045	Rest

2.2.2   The methodology and technology of phosphate coating deposition

The utilisation of phosphate coatings for the protective preservation of steel surfaces has been recognised since the advent of the 20th century, a period during which the majority of global machinery, refrigeration systems and furniture production underwent this process.The earliest recorded instance of phosphate coatings being employed for the prevention of iron and steel corrosion is evidenced by a British patent granted in 1869 to Ross. The method he employed entailed the immersion of heated iron items in phosphoric acid to achieve rust prevention. Subsequent to this early development, numerous advancements have been made in the field of phosphate coatings, contributing to their widespread utilisation in various industrial applications [1].

A variety of techniques are available for the control of corrosion, including the use of inhibitors, coatings, or by applying protection which can be anodic/cathodic [2]. Surface treatment is a very effective technique for improving corrosion resistance, including chemical conversion coating, galvanizing, physical vapor deposition, etc.[3,4]. Among these surface treatment methods, chemical conversion treatment is a simple and cost-effective method used in many applications.

Phosphating can be defined as the process of forming a layer of insoluble phosphates on the metal surface by a chemical reaction between the metal surface and the phosphating solution. This coating involves the formation of a phosphate layer that covers the entire surface of the material, as well as forming bonds with the base material [5].

In the phosphating process, surface preparation constitutes a pivotal step, as the structure and surface texture of the material frequently exert either a detrimental or beneficial influence on the phosphate layer to be deposited. To ensure the uniformity of the coating on the steel specimens (OB37), it is imperative to meticulously clean and activate their surfaces. This process can be achieved through various methods, including grinding, polishing, or by subjecting the specimens to a ceramic particle blasting [6,7].

In order to ensure a uniform coating on the surface of the material during the phosphating process, the following steps must be conducted:

- sample preparation;
- alkaline chemical degreasing;
- washing with cold running water;
- hot water washing;
- acid chemical pickling;

- washing with cold running water;
- phosphating;
- washing for neutralisation;
- double washing with running water;
- drying.

Depending on the nature of the substrate to be treated with phosphating, and also depending on the constituents of the solutions employed, certain of the aforementioned steps may be eliminated or other steps may be introduced.

In order to enhance the corrosion resistance of the phosphate coating and ensure optimal adhesion to the OB37 bar, it is essential to meticulously prepare the bar's surface. The OB37 bars were precisely cut into samples with a diameter of 10 mm and a thickness of 3 mm. These samples were then methodically grinded using SiC sandpaper with grit sizes of 400, 600, 800, 1000 and 1200, to ensure a smooth and even surface texture.

Following the preparation of a homogeneous surface, the samples were subjected to a degreasing process in an ultrasonic bath containing a mixture of ethyl alcohol and distilled water for a duration of 10 minutes, subsequently followed by an additional 15 minutes. Thereafter, a phosphate layer was deposited onto the surface of the samples by immersing them in a phosphating solution. The final stage of the phosphating process entails drying, which can be executed either at ambient temperature or at elevated temperatures ranging from 100 to 150 degrees Celsius using ovens [8-10].

As illustrated in Fig. 2.2, the installation can be found in the Department of Materials Engineering and Industrial Safety, which is part of the Faculty of Materials Science and Engineering at Gheorghe Asachi Technical University of Iasi. The installation has been utilised in the various stages of the phosphating process.

The degreasing and phosphating solutions are heated in thermostated digital baths of the DIGIBATH-2 Raypa type, thus maintaining the working temperature. The stirring of the degreasing and pickling baths is accomplished by two stirrers driven by electric motors of the SIEMES 1AF 2210 0A, 220V type, while the stirring of the phosphating solution is performed by a Heidolph R2120 stirrer at a speed of 500 rpm. Dried samples are then subjected to crystalline chemical phosphating, following which they are placed in an oven model APT.lineTM ED (E2) that can be set to a temperature range of 100-150 °C.

It is a well-established fact that steel possesses a low level of corrosion resistance, which leads to its oxidation when exposed to moisture. To mitigate this issue, the steel undergoes a coating process during the manufacturing stage, and this process is particularly important during storage. The coating applied to steel consists of a thin layer of oil or grease, which acts as a barrier against moisture and corrosion.

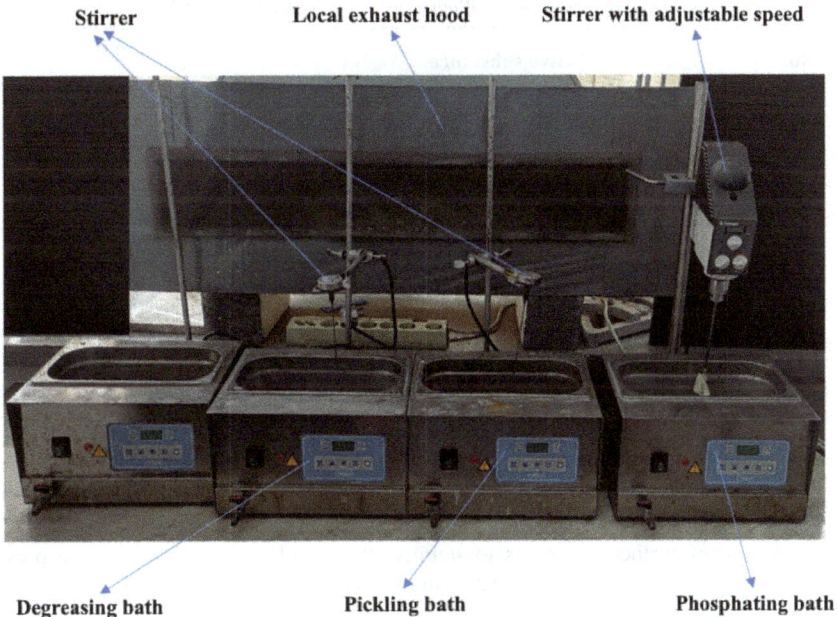

Figure 2.2.   General view of the phosphating equipment.

The presence of an oil film is known to hinder the adhesion of the phosphate layer to the substrate, underscoring the significance of degreasing as a pivotal step in the phosphating procedure. The fundamental objective of stripping is to eradicate this oil film and other residual substances that impede the activation or partial activation of the surface.

The duration of the stripping stage is not predetermined; it is conducted in accordance with the degree of contamination present in the sample. To assess the surface condition, the sample can be periodically removed from the pickling solution and examined.

A method of ascertaining the completion of the pickling process is to observe a uniform film on the surface of the sample upon its removal from the water.

The chemical alkaline degreasing solution , which was utilised in the thesis, has the chemical composition shown in Table 2.3.

Table 2.3.   Chemical composition of the degreasing solution used.

No.	Active substance	Quantity * [g]
1.	Sodium carbonate (Na_2CO_3)	60
2.	Trisodium phosphate ($Na_3PO_4 \cdot 10H_2O$)	60
3.	Sodium hydroxide (NaOH)	90
4.	Sodium silicate ($Na_2SiO_3 \cdot 9H_2O$)	10
5.	Surfactant	13

Sodium hydroxide: was first employed in the manufacture of soap, and it is now the principal constituent of pickling solutions. It is utilised as a base in the chemical industry, where it is employed in the production of sodium salts and detergents, as well as for pH regulation and organic synthesis [11].

The dissolution of solid sodium hydroxide in water is a strongly exothermic reaction, which results in the release of a significant quantity of heat. The resulting solution is typically colourless and odourless.

The addition of surfactants serves to stabilise the dissolved substances and to prevent redeposition of fat particles in the degreasing solution.

Sodium carbonate: an active component commonly found in detergent powders and degreasing solutions, is highly effective in removing oil, lubricants and grease from metal surfaces [12].

In a 200 ml distilled water solution, 60 g of sodium carbonate was dissolved, resulting in an exothermic reaction. The resulting solution is an alkaline solution, comprising carbonate anions and hydroxyl groups.

Sodium silicate: is the generic designation for compounds with the chemical formula $Na_2SiO_2 + x$. Sodium metasilicate was utilised as a degreasing solution and is a solid, crystalline, white, hygroscopic chemical compound with the chemical formula Na_2SiO_3 [13].

The surfactant (detergent): constitutes a principal component of the degreasing solution. Upon dissolution in a solution or water, its particles migrate towards the interface between liquid and solid (dirt), thereby altering its properties and facilitating the removal of dirt particles from the metal surface (www.essind.com). For the degreasing solution, 13 g of commercially available surface-active detergent was utilised, which was initially dissolved in 66 ml of distilled water.

The components of the solution were mixed, and the resulting 2000 ml of degreasing solution was supplemented with distilled water.

Scouring is a preliminary step to the phosphating operation, during which pickling is employed to clean the surface of the sample, thus removing acids, fats or inorganic substances from the metal surface.

The main purpose of pickling when preparing the surface for phosphating is to remove the oxide layer that forms. Although this can be achieved by sandblasting, salt baths or brushing, pickling is the most commonly used method today and has a number of advantages, including: it evens out the surface and increases the life of the part.

There are a number of factors to consider when it comes to pickling:

- The type of material to be pickled;
- The concentration of ferrous chlorides in the solution;
- The adhesion of the oxides;
- The acid concentration in the solution;
- The presence of inhibitors;
- The temperature at which the pickling is performed;
- Agitation (if the installation in which the process is performed is equipped with an agitator);
- The duration of immersing the sample in the pickling solution.

For OB37 steel samples, pickling shall be performed after the degreasing step. The samples shall be immersed in the pickling solution for 20 minutes. The chemical composition of the pickling solution is given in Table 2.4.

Table 2.4.   Chemical composition of the pickling solution used.

Name of active substance	Quantity *
Hydrochloric acid (HCl)	300 ml
Hexamethylenetetramine ($C_6H_{12}N_4$)	0.9 g
Sodium sulphate (Na_2SO_4)	0.3 g

Hexamethylenetetramine, also known as methanamine, hexamine or its trade name Urotropin, is a heterocyclic organic compound with the formula $(CH_2)_6N_4$. This white crystalline compound is very soluble in water and polar organic solvents. It is similar in structure to adamantane. It is useful in the synthesis of other organic compounds including plastics, pharmaceuticals and rubber additives. Its direct transition from the solid to the gaseous state, without passing through the liquid state, is achieved in a vacuum at 280°C.

Hexamethylenetetramine inhibits corrosion and, by adsorption to the metal surface, leads to electron yield. It is an inhibitor that blocks the active sites and speeds up the absorption process, reducing the corrosion rate and increasing the life of the sample in contact with the pickling solution [14,15].

Hexamethylenetetramine (0.9 g) was dissolved in 33 ml of distilled water, and then added to the hydrochloric acid contained within the pickling solution.

Sodium sulfate (Na$_2$SO$_4$) is an anorganic compound. It is found in solid form and is highly soluble in water.

The substitution of sodium sulfate for sulphuric acid in the pickling solution is driven by environmental concerns, as sulphuric acid is recognised as a highly environmentally aggressive acid [16].

The pickling solution was prepared by dissolving 0.3 g of sodium sulfate in 33 ml of distilled water.

Hydrochloric acid is defined as an aqueous solution of hydrogen chloride, otherwise known as hydrogen chloride (HCl). This solution is classified as a hard anorganic acid. The salts derived from hydrochloric acid are classified as chlorides, with sodium chloride (NaCl) being the most commonly known.

In the pickling solution, 300 millilitres of hydrochloric acid (33% concentration), 0,3 grams of sodium sulphate previously dissolved in distilled water and 0,9 grams of hexamethylenetetramine dissolved in 33 millilitres of distilled water were used, and the solution was made up to 2 litres with distilled water.

Following the degreasing and pickling stages, the samples were rinsed using either a water jet or by immersing them in water. Following the processes of degreasing and pickling, the OB37 steel samples were subjected to an immersion in distilled water, in order to remove the chemical compounds present on the surface of the samples.

Following a thorough cleaning of the samples' surfaces, the phosphating step was initiated. This process results in the formation of an insoluble phosphate layer, which provides corrosion resistance, on the surface of the samples.

Phosphating solutions can be categorised according to the nature of the metal ions that constitute the predominant component of the phosphating solution as follows: zinc, manganese or iron phosphate baths. The selection of these solutions is made with consideration for both the material to be phosphated and the properties to be obtained following deposition.

In order to obtain a phosphate coating that will improve the corrosion resistance of OB37 steel, as well as a coating that can be used as a substrate for future coatings, the solutions presented in Tables 2.5, 2.6 and 2.7 were used. The quantities of active substances utilised were calculated for a 2-litre phosphating solution, to which bidistilled water was added.

Table 2.5.   Chemical composition of I/Mg phosphating solution.

Active substance	Quantity *
Sodium hydroxide (NaOH)	7 g
Sodium nitrite (NaNO$_2$)	0.4 g
Magnesium carbonate (MgCO$_3$)	8.5 g
Phosphoric acid (H$_3$PO$_4$) 85%	23 ml

* quantities are given for 2l of solution.

Table 2.6.   Chemical composition of the II/Zn phosphating solution.

Active substance	Quantity*
Sodium hydroxide (NaOH)	0.9 g
Sodium nitrite (NaNO$_2$)	0.6 g
Sodium tripolyphosphate (Na$_5$P$_3$O$_{10}$)	0.1 g
Phosphoric acid (H$_3$PO$_4$)	22.00 ml
Nitric acid (HNO$_3$)	11.00 ml
Zinc (Zn)	9.00 g

* quantities are given for 2l of solution.

Table 2.7.   Chemical composition of III/Mn phosphating solution.

Active substance	Quantity *
Sodium hydroxide (NaOH)	0.75 g
Sodium nitrite (NaNO$_2$)	0.45 g
Sodium tripolyphosphate (Na$_5$P$_3$O$_{10}$)	0.05 g
Phosphoric acid (H$_2$PO$_4$)	7.00 ml
Nitric acid (HNO$_3$)	0.40 ml
Nickel (Ni)	0.03 g
Iron (Fe)	0.03 g
Manganese (Mn)	1.50 g

* quantities are given for 2l of solution.

The phosphating solutions under consideration contain the following accelerators and inhibitors: HNO_3, $NaOH$, $NaNO_2$, and $Na_5P_3O_{10}$, in varying proportions [8-10]. The phosphating step was executed for a duration of 60 minutes at a temperature of 90°C. The addition of $MgCO_3$, Zn chips, Ni, Fe, Mn powder, and H_3PO_4 to these solutions was essential for the formation of metal compounds, which subsequently led to the establishment of the phosphate layer [17]. Subsequent to this step, the samples were rinsed with water and dried at room temperature [18].

The sequence of steps in the phosphating process is illustrated in Fig. 2.3.

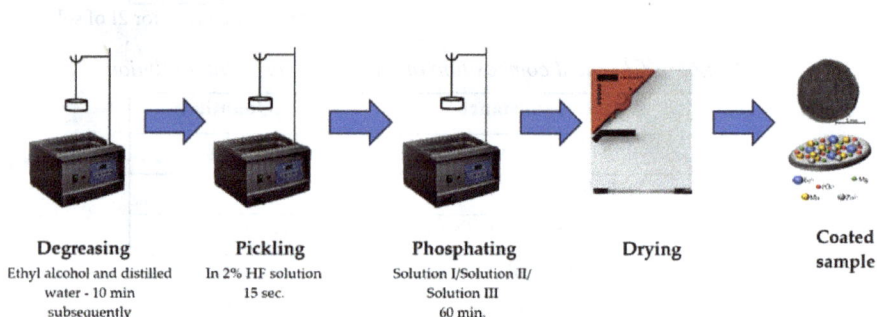

Degreasing	Pickling	Phosphating	Drying	Coated sample
Ethyl alcohol and distilled water - 10 min subsequently	In 2% HF solution 15 sec.	Solution I/Solution II/ Solution III 60 min.		

Figure 2.3.   The sequence of steps in the phosphating process.

The feasibility of applying phosphate coatings to metal surfaces is contingent upon the dimensions and configuration of the substrate, in addition to the intended application. The method of application, whether by spraying or dipping, is determined by the specific characteristics of the coating and the intended use of the metal surface. The selection of the most suitable method is determined by the phosphater's preference.

Immersion phosphating is a method that allows for the uniform coating of metal surfaces. However, this method is disadvantageous due to the extended duration of the immersion process. Given the focus of this thesis on OB37 steel for concrete reinforcement and the relatively small size of the samples, the immersion method was selected for the phosphating process. This method was applied on the phosphating equipment shown in Fig. 2.2, and the succession of the phosphating process steps followed the steps shown in Fig. 2.3.

Following the phosphating step, it is imperative to rinse the samples to eliminate soluble active salts. The rinsing of the phosphated samples was proceeded with distilled water.

The final stage of the phosphating process is the drying step. Drying can be accomplished by evaporation, by air currents, or by heating the samples. In this study, the drying process was executed using a Binder FD 23 drying oven, which employs forced convection at a temperature of 300°C and a capacity of 20 liters (Fig. 2.4).

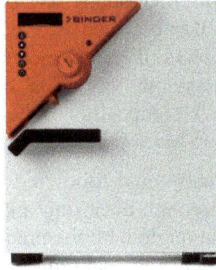

Figure 2.4.   Binder FD 23 drying oven overview.

2.3   Experimental details of analysis methods and the equipment used

The obtained phosphate layers were characterized in terms of structural, mechanical, and corrosion resistance, as well as reinforcement-concrete adhesion. The samples utilized in this study were obtained using an electroerosion cutting machine, EDM DEM 320A 400A-500A (Huayuan Road, Haidian District, Beijing, China). The samples used in the experimental research were those of OB37 steel used for reinforcement of concrete structures.

2.3.1   Optical microscopy

The structural integrity of a metallic material is contingent upon its chemical composition, as well as the mechanical processing and heat treatment to which it is subjected. The material's structural characteristics are contingent on its state, exhibiting distinct structures in different states. The structural characteristics of a material can be determined by means of fast and inexpensive physical analysis methods. Direct, indirect, and auxiliary methods can be employed [19].

Optical microscopy analysis provides images of the morphology and size of the crystals on the surface of the phosphate layer, with magnification powers ranging from 50X to 1000X.

Optical microscopy constitutes the method of analyzing the structure of materials where light is directed vertically through the microscope objective and reflected back through the objective to an eyepiece, viewing screen, or camera.

2.3.2   Scanning electron microscopy (SEM)

Scanning electron microscopy (SEM) is a method for high-resolution imaging of surfaces. This technique utilizes electrons for imaging, similar to how an optical microscope uses visible light. In comparison to optical microscopy, it offers a magnified image with a resolution up to 50000X, while achieving a field depth up to 100X.

In the process of scanning electron microscopy, the surface of the specimen is scanned with an electron beam. The interaction between the electron beam and the material results in the emission of electrons and photons as the electrons penetrate the sample surface. The resulting particles are subsequently collected by a detector, thereby furnishing information regarding the surface's composition. The image of the surface is thus the result of the collision of the electron beam with the surface of the sample being analyzed.

The sanding was conducted manually, and following each transition in the selection of paper, the sample underwent a thorough cleansing and a 90° rotation to ensure that the newly formed traces formed a right angle with the preceding ones, thereby preserving the flatness of the sample.The transition to the subsequent metallographic paper occurred as soon as it was determined that all the traces from the previous sanding had been eradicated. To avoid overheating the samples during high-speed sanding, a cooling system integrated within the equipment was utilised.

The objective of the metallographic sample polishing procedure is to achieve a surface with a "mirror gloss". The polishing was conducted mechanically using a polishing machine equipped with a rotating disk to which a felt impregnated with aluminium oxide had been attached.

Following this procedure, the samples were thoroughly rinsed with water to remove any residual aluminium oxide, then degreased using alcohol, and finally dried with filter paper.

The polish was examined under an optical microscope at 100X magnification.

It is imperative that the coating process be subject to rigorous quality control measures to ensure optimal coating performance.The quality of the conversion coatings and the analysis of the crystal morphology were conducted using the electron scanning microscope Vega Tescan LMH II, which is part of the Department of Materials Science at the Faculty of Materials Science and Engineering, at the "Gheorghe Asachi" Technical University of Iasi.

2.3.3 Energy dispersive X-ray spectroscopy (EDX)

Energy-dispersive X-ray spectroscopy (EDX) is utilised in conjunction with scanning electron microscopy (SEM), facilitating the identification and quantification of chemical compositions. Consequently, the OB37 steel samples were analysed from this perspective following phosphating with three distinct solutions. This approach facilitated the determination of the chemical composition of the deposited layers, their concentration, and the existence of any impurities.

The chemical composition of the phosphate layers deposited on the surface of the OB37 steel samples was determined by utilising the Bruker EDAX detector attached to the electron scanning microscope (model Vega Tescan LMH II). This microscope serves as a valuable instrument within the Materials Science Department of the Faculty of Materials Science and Engineering at the "Gheorghe Asachi" Technical University of Iasi.

2.3.4   X-ray diffraction (XRD)

X-ray diffraction is an analytical technique used to characterize solid, organic, or inorganic materials [20]. The diffraction intensities are generated as a result of the interactions of X-rays with the electrons of atoms in the lattice.

The specific diffractometric technique employed hinges on the manner in which the compounds under investigation manifest themselves, with powder diffraction being utilised for crystalline powders and single crystal diffraction. The structural characterisation of the crystal lattice is facilitated by the information extracted by these two diffractometric techniques [20].

X-ray diffraction is a method of obtaining information with great rapidity. It is a non-destructive method, and only a small amount of sample is required. The following information can be extracted from X-ray powder diffractograms:

- Qualitative crystal phase analysis: based on each crystal phase having a specific diffractogram;

- Quantitative phase analysis is a process that enables the determination of the percentage value of each phase identified in a sample, provided that the sample contains multiple phases. This is achieved by calculating the number and intensity of the diffraction lines present;

- The process of microstructural analysis encompasses the determination of crystallite sizes, lattice stresses and defect probabilities;

- The degree of crystallinity can be determined through the calculation of the ratio of the area of the diffraction maxima to the total area. The total area includes both of these areas, as well as the area of the halos that are indicative of the amorphous phase;

- Determination of the crystalline structure of powders: single-crystal X-ray diffraction is the technique used to obtain the crystal structure of the compound under investigation. [20].

X-ray diffraction analysis was conducted to ascertain the nature of the crystals that formed on the surface of phosphated OB37 steel. This analysis was performed using an X-ray diffractometer (model PANalytical X'Pert PRO MPD) located in the Advanced Materials Laboratory at the Faculty of Mechanics, "Gheorghe Asachi" Technical University of Iasi.

2.3.5   Mechanical properties determination by scratch and microindentation tests

The scratch test is a simple and rapid method for characterising coatings. In this study, the adhesion of the zinc phosphate coating was determined, as well as their coefficient of friction.

The test method involves subjecting the sample to a continuously increasing force while it is moved at a constant speed. It is imperative to ensure that the thickness of the coating, the mechanical properties of the substrate, and the test conditions are carefully considered, as they have the capacity to influence the test results [21].

The scratch test is a method of assessing the mechanical properties of materials by means of a controlled abrasion process [16]. The test parameters are as follows: a microblade is moved at a constant speed of 167 µm/s across a distance of 10 mm over a period of 60 seconds. The material is progressively removed from the sample surface. The test quantifies several parameters, including F_x, representing the friction force value, F_z, representing the normal force, and COF, representing the coefficient of friction [9]. In this test, the normal force increases with time, ranging from 0 to 10 N.

The microindentation test is a method used for determining the mechanical properties of metals due to the ease and speed with which it can be performed [22,23]. It has been utilised to evaluate changes in the mechanical properties of deposited coatings, including hardness and modulus of elasticity.

The micro-indentation test is comprised of two primary stages: an initial pre-loading stage, with a force of 1% of the maximum test force, and a loading stage, with a force of up to 450 N, depending on the set value and the material being tested.The purpose of the pre-loading stage is to establish the 0 reference point of the graph, which is followed by an increase in force, constant over time, up to the maximum value set. Once this value is reached, the loading force is held constant for a brief period. This is followed by a decrease in the loading force, known as unloading, which is accomplished at a constant rate over a predefined period of time [24].

The hardness, as measured by the indentation process, is defined as the average contact pressure, denoted H_{IT} and is given by the following equation:

$$H_{IT=}F_{max} \ / \ AP \qquad\qquad (2.1.)$$

where F_{max} is the maximum force and A_P is the projection of the contact area at this force.

The characteristic curves of indentation tests are obtained by plotting the variation of the loading force, measured in Newtons (N), and the material strain, measured in micrometers (µm), against the variation of the unloading force.The irregularities exhibited by the loading curve are indicative of the presence of cracks in the tested material when subjected to force [22,23]. In certain instances, irregularities may also manifest on the unloading curve, attributable to the adhesion of the material to the indenter.The distance plotted on the abscissa, between the 0 point and the end point of the test, signifies the residual deformation of the material post-test [25,26].

The determination of the mechanical properties of the deposited phosphate and paint layers was conducted utilising the universal micro-tribometer CETR UMT-2 from the Tribology Laboratory of the Faculty of Mechanics at the "Gheorghe Asachi" Technical University of Iasi.

Materials Research Foundations **183** (2025)   https://doi.org/10.21741/9781644903810

2.3.6   Corrosion resistance of phosphate coating

Corrosion can be defined as the spontaneous and irreversible physico-chemical destruction of metals or alloys under the chemical, electrochemical or biological action of the environment.

The nature of the metal and the corrosive environment are the primary factors that influence the corrosion process. Other variables that must be considered include pressure, temperature, and the static or dynamic nature of the corrosive environment.

It is widely accepted that the majority of corrosion processes in natural environments are driven by electrochemistry. Given this, the electrochemical study of corrosion has emerged as a promising approach to gaining insights into the thermodynamic probability of metal corrosion in a liquid, the instantaneous corrosion rate (defined as the corrosion rate upon direct immersion of the metal in the corrosion medium), the nature of corrosion, and the factors that can influence corrosion (such as temperature, pH, and the presence of corrosion accelerators or inhibitors).

The pH value of the solutions was determined using a pH meter type RADELKIS OP-264/1 RADELKIS OP-264/1 at the Faculty of Chemical Engineering and Environmental Protection "Cristofor Simionescu", "Gheorghe Asachi" Technical University, Iasi.

The corrosion behaviour of all samples was analysed by linear polarisation and cyclic voltammetry. The equipment used was an OrigaFlex OGF+01A potentiostat (OrigaLys ElectroChem SAS, Rillieux-la-Pape, France) with a three-electrode electrochemical cell consisting of a platinum auxiliary electrode, a calomel reference electrode and a working electrode. Sample sizes of 7 mm diameter and 3 mm thickness (exposed surface area $0.384 \ cm^2$) were used. The potentiodynamic curves were recorded with a scanning rate of 0.5 mV/s and a potential range of $-350 \div +400$ mV. For cyclic voltammetry tests, the initial potential was -400 mV and the final potential was +500 mV, with a scanning rate of 10 mV. The curves obtained were recorded after 60 minutes of immersion in the solution (river water or sea water, respectively).

Electrochemical impedance spectroscopy (EIS) studies were performed on OB37 samples phosphated with phosphating solution II/Zn and OB37 control samples immersed in river water and seawater for one hour, using the OrigaFLEX OGF+01A EIS OrigaFLEX OGF+01A potentiometer (OrigaLys ElectroChem SAS, Rillieux-la-Pape, France) and experimental data acquisition was performed using OrigaMaster software (version 5, OrigaLys ElectroChem SAS, Rillieux-la-Pape, France). The experimental data were processed using the ZSimpWin software (version 3.5, E-chem Software, Ann Arbor, MI, USA), in which the spectra were fitted using the Boukamp method.

2.3.7 *Maximum tensile strength determination and evaluation of concrete-rebar adhesion*

The test equipment

The adhesion test of the steel reinforcement was performed according to SR EN 10080:2005 in conjunction with information from the scientific literature. The following equipment, shown in Fig. 2.5, was used to perform the adhesion test.

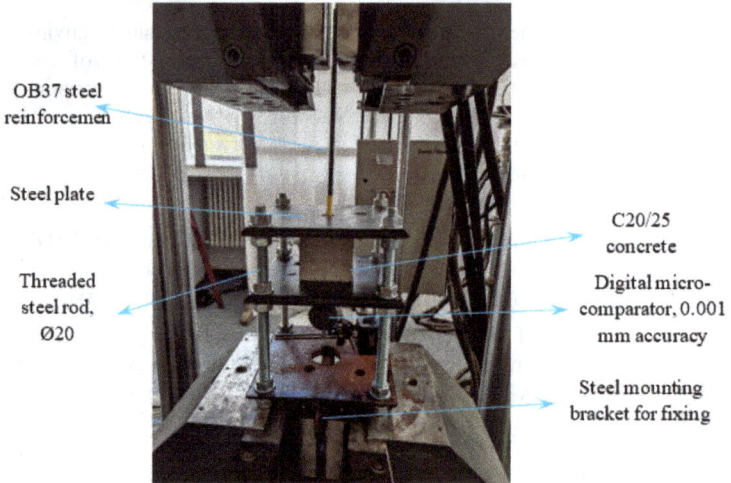

Figure 2.5.   Adhesion test equipment.

The fixture consists of 3 18 mm thick steel plates connected by 4 threaded steel rods of 20 mm diameter. The position of the central steel plate can be adjusted to accommodate different sizes of concrete samples. The top plate has a central circular recess of 20mm diameter for testing different diameters of rebar.

Both the central and lower steel plates of the fixture are provided with a 50mm diameter central circular hole to facilitate the use of different types of measurement instruments.

The main requirement for the device is that it must be rigid so as not to introduce any additional deformation that could affect the accuracy of the data recorded during the experiments.

Materials

Smooth mild steel reinforcement, OB37, was used for the adhesion tests. The treatment of the steel reinforcement with different phosphating solutions was the main parameter of the experimental design. The choice of smooth reinforcement is justified by the fact that it was not desired to introduce additional approximations in the adhesion calculation (approximation of the surface of the reinforcement with striations, etc.).

The steel used for the reinforcement had a yield strength of fy = 235 MPa. As the quality of the steel was guaranteed by the manufacturer, it was not considered necessary to conduct tensile tests on the reinforcement coupons.

The reinforcement was embedded in C20/25 concrete. This is one of the most widely used concrete classes in the construction industry. The concrete mix design is shown in Table 2.8.

Table 2.8.   Concrete mix design class C20/25 (1 mc)

Cement	Aggregate			Water	Superplasticizer	Water / cement
	0-4 mm	4-8 mm	8-16 mm			
[kg/m^3]	[kg/m^3]	[kg/m^3]	[kg/m^3]	[L]	[L]	-
460	520	361	834	176.64	2.76	0.39

A SikaPlast 331 superplasticiser (water-reducing agent) was used to maintain a low water/cement ratio. The aggregates were used in a saturated state so as not to absorb any of the water required to hydrate the cement. The cement used was CEM II A-LL 42.5R, a quick-setting composite cement manufactured in accordance with SR EN 197-1:2011.

Specimen preparation

As mentioned above, the main parameter of the experimental program was the surface preparation of the reinforcement in the concrete embedment area using 3 different phosphating solutions. In addition, specimens with untreated steel reinforcement were also prepared as control samples.

For each of the 4 cases, 6 specimens were produced. The wooden formwork was impregnated with oil so as not to absorb water from the fresh concrete and not to affect the hydration of the cement, which would have resulted in a lower quality of concrete than originally selected. Before the samples were poured into the formwork, Fig. 2.6, the formwork was additionally treated with a Sika Separol release agent.

Dry mixing was chosen for concrete preparation. The cement was first mixed with the sand for 3 minutes. Additional measures were taken to avoid losing the fine part of the powder during mixing by covering the mouth of the concrete mixer with foil. Then 75% of the water previously mixed with the superplasticiser was added. Mixing continued until no more cement deposits mixed with sand were observed on the walls of the concrete mixer. The 4-8 mm and 8-16 mm aggregates were added one at a time with the remaining water from the concrete mix design and mixing continued for a further 5 minutes.

All specimens subjected to the adhesion test were cast from the same concrete slurry, Fig. 2.7, to eliminate any possible variation due to concrete mixing. In addition, 10 cylindrical specimens measuring 100 mm × 200 mm (diameter × height) were cast to determine the mechanical properties of the 28-day concrete.

Immediately after casting and vibrating, the concrete specimens were covered with plastic film to retain moisture on the surface of the specimens and prevent contraction cracking, and stored for 24 hours.

Figure 2.6.   Formwork preparation for specimen casting.

Figure 2.7.   Adhesion test specimens.

The specimens were stripped 24 hours after casting and stored in water, Fig. 2.8, for further hydration until 28 days. The steel reinforcement was protected with self-adhesive waterproof tape at a distance of 5 cm from the surface of the concrete cube to prevent corrosion.

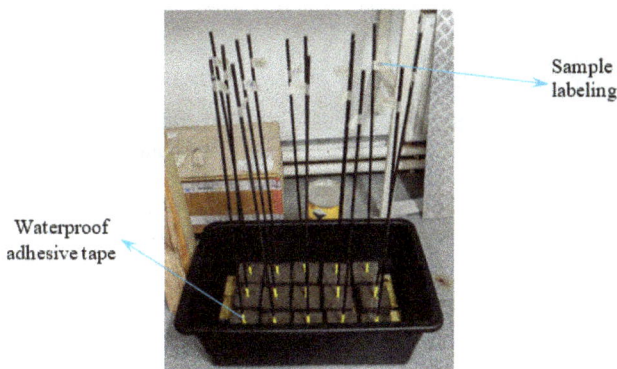

Figure 1.   Sample storage up to 28 days.

2.4   Conclusions

Because of its use in construction, OB37 steel has to be as durable as possible in every respect. Reinforcement is the central part of a structure that must last for a very long time. The immersion phosphating process is a relatively simple and inexpensive method of obtaining phosphate layers that improve the properties of OB37 steel. The main property that is improved is corrosion resistance. This increases the range of applications for this steel.

Using phosphatised coatings increases the porosity of the surface, thereby improving or maintaining adhesion, resulting in better concrete-reinforcement adhesion. An improved porosity is also a great help for other types of coatings or can be used as a base for them.

To demonstrate that the phosphatization of OB37 rebar, through the deposition of new phosphate layers of three different compositions on its surface, makes a significant contribution to its properties, modern equipment has been used to characterize them from a chemical, structural and also mechanical standpoint.

References

[1]   Hafiz M.H., Kashan J.S., Kani A.S., Effect of Zinc Phosphating on Corrosion Control for Carbon Steel Sheets, 2008, Eng. & Technology, 26(5), 501-511.

[2]   Jalili M.M., Moradian S., Hosseinpour D., The use of inorganic conversion coatings to enhance the corrosion resistance of reinforcement and the bond strength at the rebar/concrete, 2009, Construction and Building Materials 23, 233–238, https://doi.org/10.1016/j.conbuildmat.2007.12.011

[3]   Ogle K., Tomandl A., Meddahi N., Wolpers M., The alkaline stability of phosphate coatings I: ICP atomic emission spectroelectrochemistry, Corrosion

Science, 46, 2004, 979–995.

[4]   Callister Jr. W.D., Materials Science and Engineering – An Introduction, seventh ed., John Wiley & Sons, Inc., 2007, ISBN-13: 978-0-471-73696-7.

[5]   Bejinariu C, Lazar P, Sandu AV, Cazac AM, Sandu IG, Corbu O, Enhancing Properties of Reinforcing Steel by Chemical Phosphatation. The ICAMET 2014, 3th International Conference Proceedings – Advanced Materials Engineering & Technology, Ho Chi Minh City, Vietnam, December 4-5, 2014, Applied Mechanics and Materials Vols. 754-755 (2015) pp 310-314, (2015) Trans Tech Publications, Switzerland, https://doi.org/10.4028/www.scientific.net/AMM.754-755.310, ISBN: 978-3-03835-434-5

[6]   Lazar, P., Bejinariu, C., Cazac, A.M., Sandu, A.V., Bernevig, M.A., Burduhos-Nergis, D.P. Phosphate coatings for the protection of steels reinforcement for concrete. (2021) Journal of Physics: Conference Series, 1960 (1), art. no. 012013. https://doi.org/10.1088/1742-6596/1960/1/012013

[7]   Lazar P, Bejinariu C, Sandu AV, Cazac AM, Sandu IG, Chemical Deposition of Thin Layers on Reinforcing Steel. International Conference on Innovative Research, May 14th to 15th, ICIR 2015, Iasi – Romania, Key Engineering Materials Vol 660 (2015) pp 213-218, (2015) Trans Tech Publications, Switzerland, https://doi.org/10.4028/www.scientific.net/KEM.660.213

[8]   Burduhos-Nergis D.P., Vizureanu P., Sandu A.V., Bejinariu C. Evaluation of the corrosion resistance of phosphate coatings deposited on the surface of the carbon steel used for carabiners manufacturing, 2020, Applied Sciences 10 (8), 2753.

[9]   Burduhos-Nergis D.P., Vizureanu P., Sandu A.V., Bejinariu C., Phosphate Surface Treatment for Improving the Corrosion Resistance of the C45 Carbon Steel Used in Carabiners Manufacturing, 2020, Materials 13, 3410.

[10]   Burduhos-Nergis, D.-P.; Vizureanu, P.; Sandu, A.V.; Bejinariu, C. Phosphate surface treatment for improving the corrosion resistance of the c45 carbon steel used in carabiners manufacturing. Materials 2020, 13, 3410. https://doi.org/10.3390/ma13153410

[11]   Bowman E., J. Varney, N. Thompson, O. Moghissi, M. Gould, J. Payer, International Measures of Prevention, 2016, Application, and Economics of Corrosion Technologies Study, Houston, Texas, USA.

[12]   Hansson C.M., Mammoliti L., Hope B.B., Corrosion inhibitors in concrete. Part I: the principles, Cement & Concrete Composites Volume 28, Issue 12, December 1998, Pages 1775-1781, https://doi.org/10.1016/S0008-8846(98)00142-2

[13]   Information on http://www.chemicalbook.com

[14]   Bajat, J.; Mišković-Stanković, V.; Popić, J.; Dražić, D. Adhesion characteristics and corrosion stability of epoxy coatings electro-deposited on phosphated hot-dip

galvanized steel. Prog. Org. Coat. 2008, 63, 201–208.
https://doi.org/10.1016/j.porgcoat.2008.06.002

[15] Królikowski A., Kuziak J., Impedance study on calcium nitrite as a penetrating corrosion inhibitor for steel in concrete, Electrochim. Acta 56 (23) (2011) 7845–7853, http://dx.doi.org/10.1016/j.electacta.2011.01.069

[16] Ernens D., de Rooij M.B., Pasaribu H.R., van Riet E.J., van Haaften W.M., Schipper D.J., Mechanical characterization and single asperity scratch behaviour of dry zinc and manganese phosphate coatings, Tribology International 118 (2018) 474–483, https://doi.org/10.1016/j.triboint.2017.04.034

[17] Valanezhad, A.; Tsuru, K.; Maruta, M.; Kawachi, G.; Matsuya, S.; Ishikawa, K. Novel Ceramic Coating on Titanium with High Mechanical Properties. Bioceram. Dev. Appl. 2011, 1, D110124.

[18] Seo, Y.H. A study on improving the prediction accuracy of cold forging die life based on quantitative evaluation of phosphate film damage. Sci. Rep. 2023, 13, 16464.

[19] Simescu, F.; Idrissi, H. Corrosion behaviour in alkaline medium of zinc phosphate coated steel obtained by cathodic electro-chemical treatment. Corros. Sci. 2009, 51, 833–840. https://doi.org/10.1016/j.corsci.2009.01.010

[20] Information on http://www.itim-cj.ro.

[21] Bhaskaran R., Palaniswamy N., Rengaswamy N.S., Jayachandran M., A review of differing approaches used to estimate the cost of corrosion (and their relevance in the development of modern corrosion prevention and control strategies), 2005, Anti-Corrosion Methods and Materials 52, 29-41.

[22] Zhang F., Pan J., Recent Development of Corrosion Protection Strategy Based on Mussel Adhesive Protein, Frontiers in Materials, 2019, 6, 207.

[23] Zhang, W.; Song, X.; Gu, X.; Li, S. Tensile and fatigue behavior of corroded rebars. Constr. Build. Mater. 2012, 34, 409–417. https://doi.org/10.1016/j.conbuildmat.2012.02.071

[24] Khaleghi M., Gabe D., Richardson M., Characteristics of manganese phosphate coatings for wear-resistance applications. Wear 1979;55(2):277–87. https://doi.org/10.1016/0043-1648(79)90159-5

[25] Bastidas D.M., Criado M., Fajardo S., La Iglesia A., Bastidas J.M., Corrosion inhibition mechanism of phosphates for early-age reinforced mortar in the presence of chlorides, Cement and Concrete Composites 61 (2015) 1–6.

[26] Bastidas D.M., Criado M., La Iglesia V.M., Fajardo S., La Iglesia A., Bastidas J.M., Comparative study of three sodium phosphates as corrosion inhibitors for steel reinforcements, Cement & Concrete Composites 43 (2013) 31–38.

41

Materials Research Foundations **183** (2025)   https://doi.org/10.21741/9781644903810

CHAPTER 3

Study of the morphology and structure of phosphate coatings on OB37 steel

P. Lazar[1], A.-M. Cazac[1]*, C. Bejinariu[1,2]

[1]Faculty of Materials Science and Engineering, Gheorghe Asachi Technical University of Iasi, Romania

[2]Academy of Romanian Scientists, Ilfov 3, 050044 Bucharest, Romania

alin-marian.cazac@academic.tuiasi.ro

Abstract

The phosphate coating applied to the surface of the OB37 steel samples was obtained through an immersion process, utilising the phosphating method. In order to provide a comprehensive analysis of the layers obtained, their chemical composition, and the compounds formed after phosphating, the following investigations were conducted: firstly, the deposited layer was structurally analysed by scanning electron microscopy; secondly, EDX and X-ray diffraction analysis were used to reveal the chemical composition of the layers and the compounds formed after phosphating.

Keywords

Phosphating Morphology and Structure of Phosphate Coatings

3.1 Structural characterization of phosphate layers by scanning electron microscopy

The structural analysis of the obtained surfaces (Fig. 3.1.a-c) reveals that, in all phosphating cases examined, the surfaces are covered with partial deposits of compounds from the electrolyte solutions utilised during the phosphating process (Table 3.1).

In all three proposed deposition cases, regions can be identified between the crystals at the surface, known as intercrystalline regions. These intercrystalline regions cause the surface of the layer to become uneven, a characteristic of phosphate layers deposited by the chemical conversion process.

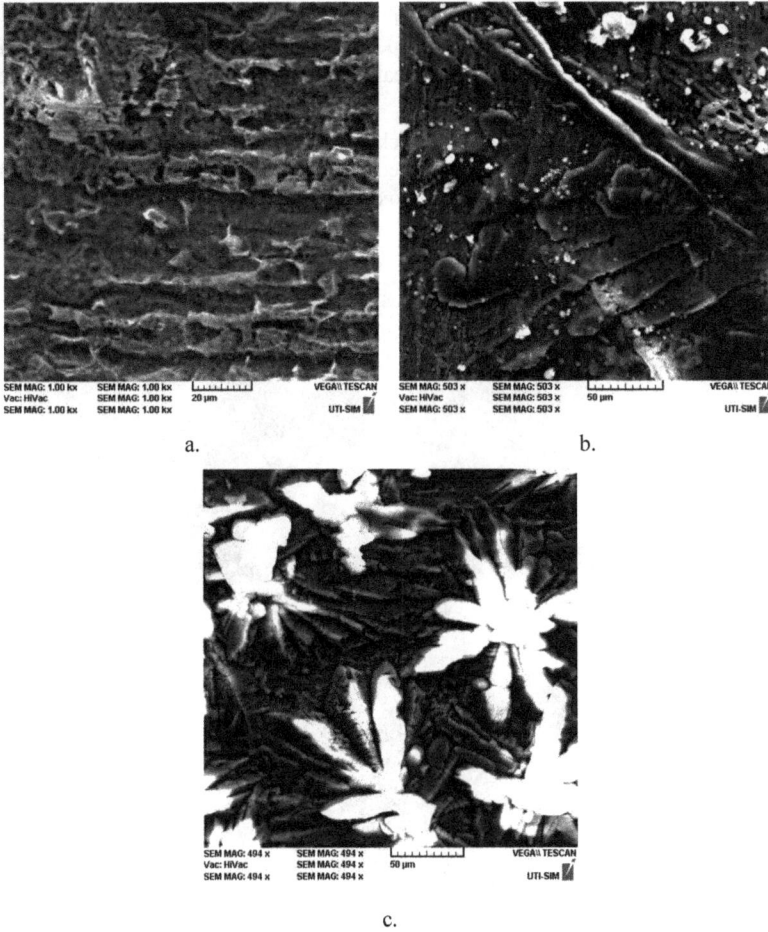

Figure 3.1.   SEM micrographs of surfaces phosphated with (a) Solution I, (b) Solution II and (c) Solution III.

The results demonstrated that the layer covering the metal substrate is not continuous, a characteristic that has been observed in the majority of layers obtained by phosphating. In all cases, an identifiable crystalline precipitate is present on the surface. The size of the crystallites varies, with the majority measuring a few micrometers in length [1].

In addition to the precipitates formed on the surface, crystals can also be observed in various forms (Fig. 3.1.a-c), which vary depending on the solution used for phosphating. The sizes of the crystallites, some cylindrical, others acicular, range from approximately 5-50 μm.

Fig. 3.2 shows SEM images of the layer thickness obtained by phosphating with the three solutions, with the thickness of the deposited layer measuring approximately 10 μm and that of the conversion layer measuring approximately 20-30 μm.

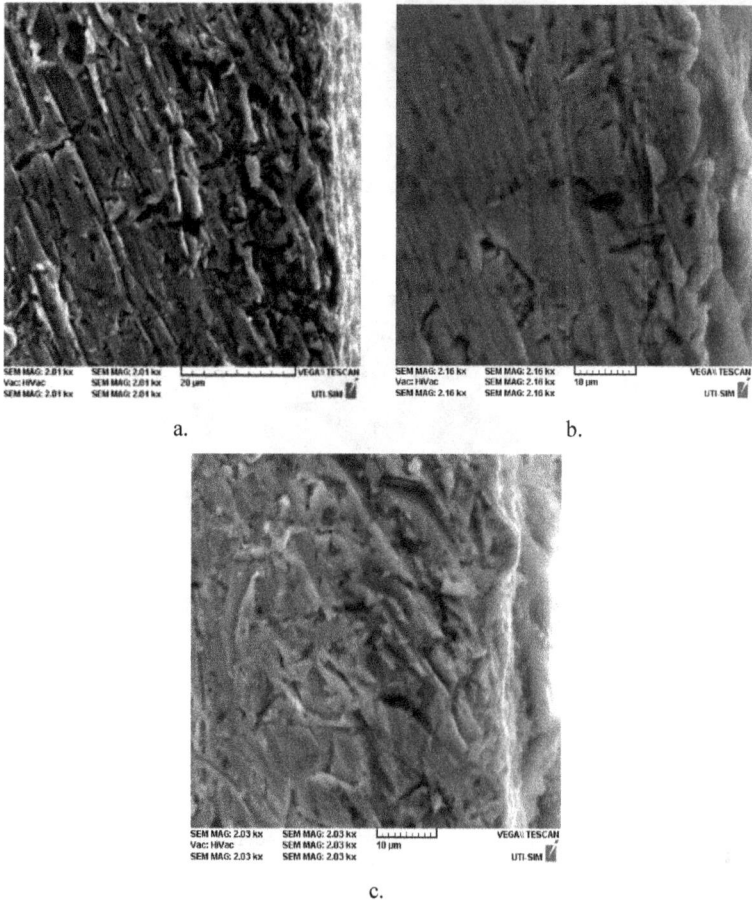

a.

b.

c.

Figure 3.2.   SEM images of the layer thickness obtained by phosphating with (a) solution I, (b) solution II and (c) solution III.

3.2   Chemical composition of phosphate coatings assessed by EDX analysis

In order to ascertain the chemical composition of phosphate layers deposited on OB37 steel samples used in concrete reinforcement, cylindrical specimens were cut from 6 mm-long bars with a diameter of 10 mm. The resulting samples had a diameter of 10 mm and a height of 3 mm.

EDX determinations were performed on the surface of each deposited layer for the three phosphating solutions, in different areas of the sample. The analysis was conducted with the Bruker EDAX detector attached to the Vega Tescan LMH II LMH II scanning microscope.

As demonstrated in Table 3.1, the surface layers obtained from phosphating procedures exhibit phosphate-based compounds in conjunction with specific chemical compounds derived from each solution.

Table 3.1.   Chemical composition of phosphate surfaces.

Surface		Phosphating solution I/Mg	Phosphating solution II/Zn	Phosphating solution III/Mn	EDS Error %
O%	wt	19.96	38.06	33.12	2.9
	at	33.89	66.55	62.01	
Fe%	wt	61.78	15.1	47.66	1.5
	at	30.05	7.56	25.56	
C%	wt	14.33	-	-	1.9
	at	32.4	-	-	
P%	wt	3.05	12.31	6.4	0.25
	at	2.68	11.12	6.2	
Mg%	wt	0.88	-	-	0.12
	at	0.98	-	-	
Mn%	wt	-	-	3.6	0.11
	at	-	-	1.96	
Ni%	wt	-	-	0.58	0.07
	at	-	-	0.3	
Zn%	wt	-	34.52	-	0.2
	at	-	14.76	-	

* St. Dev: O: ± 2.1; Fe: ± 1.1; C: ± 1.1; ± 2.1; P: ± 0.7; Mg: ± 0.1; Mn: ± 0.15; Ni: ± 0.05; Zn: ± 1.2.

In accordance with the findings of preceding studies [2], one of the compounds formed on the surface is phosphophyllite. For the second solution, the presence of $Zn_2Fe(PO_4)_24H_2O$ phase was confirmed [3]. Additionally, for the Zn-phosphated sample, solution II, the presence of zinc phosphate tetrahydrate was observed (Fig. 3.3).

The introduction of Mn into the phosphate solution resulted in the formation of another phase, namely $Mn2.5(HPO_4)(PO_4)(H_2O)_2$ [3]. In the context of manganese phosphating, the addition of nickel to the phosphating process serves the function of a catalyst, thereby accelerating the formation of manganese phosphate crystals. The addition of Ni to the phosphating solution has been shown to facilitate the formation of a sealed layer of ordinary rhombic crystals, as illustrated in Fig. 3.1.c. The predominant Ni compound observed was nickel phosphate, as depicted in Figure 3.5.

Fig. 3.3., 3.4. and 3.5. show the EDX spectra of the obtained phosphate layers and their EDX microstructure.

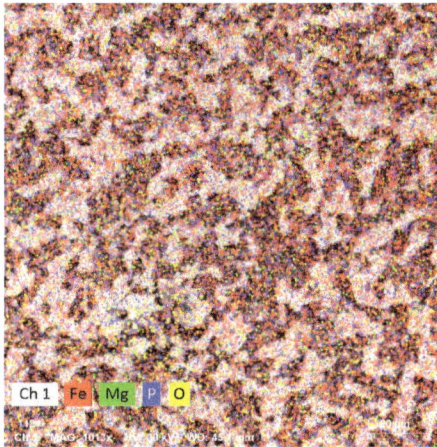

a.

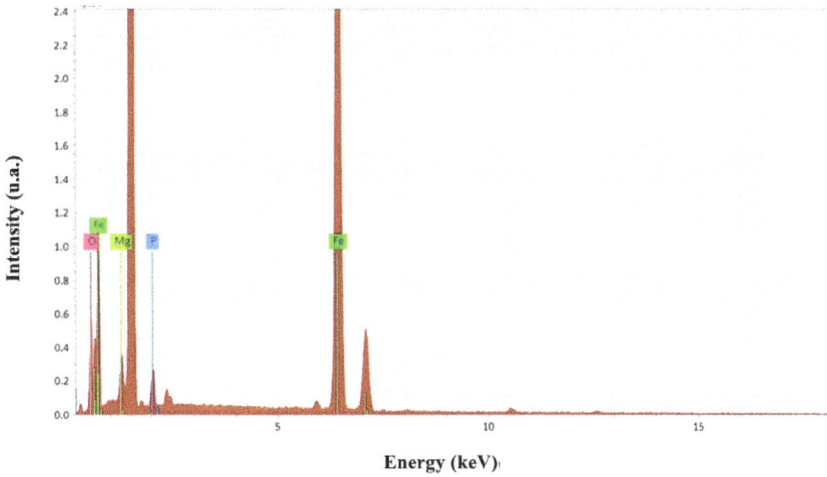

b.

Figure 3.3.   EDX analysis of the sample phosphated with I/Mg phosphating solution, (a) elemental chemical distribution, (b) EDX spectrum.

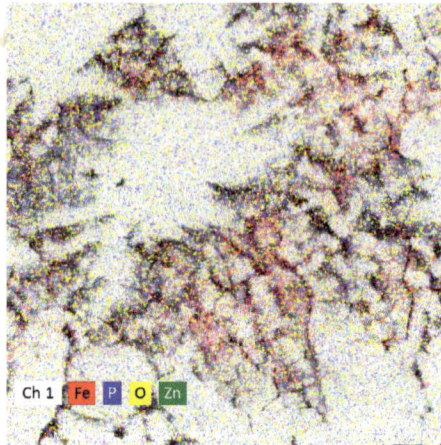

a.

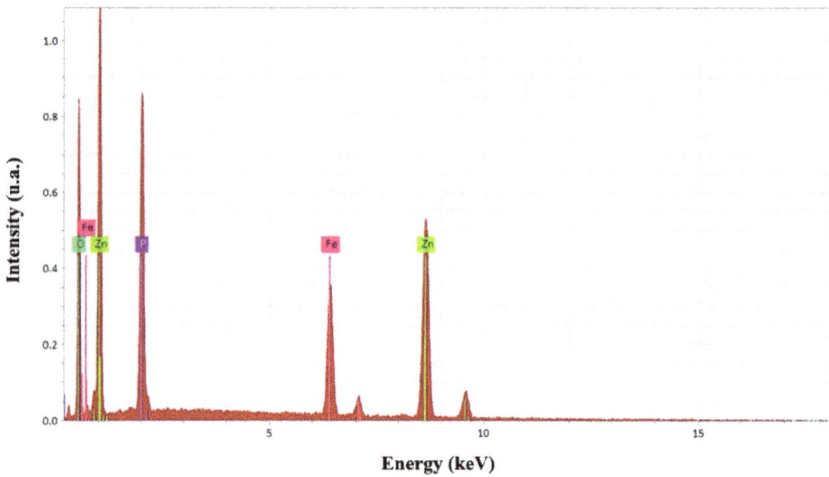

b.

Figure 3.4.   EDX analysis of the sample phosphated with II/Zn phosphating solution, (a) elemental chemical distribution, (b) EDX spectrum.

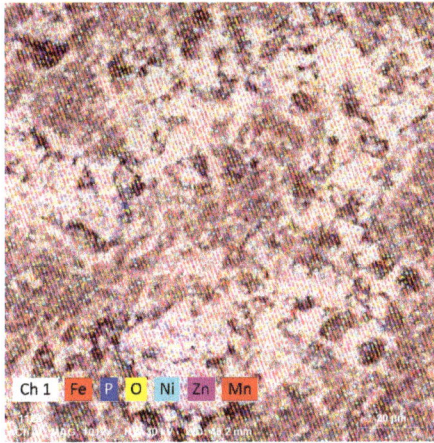

a.

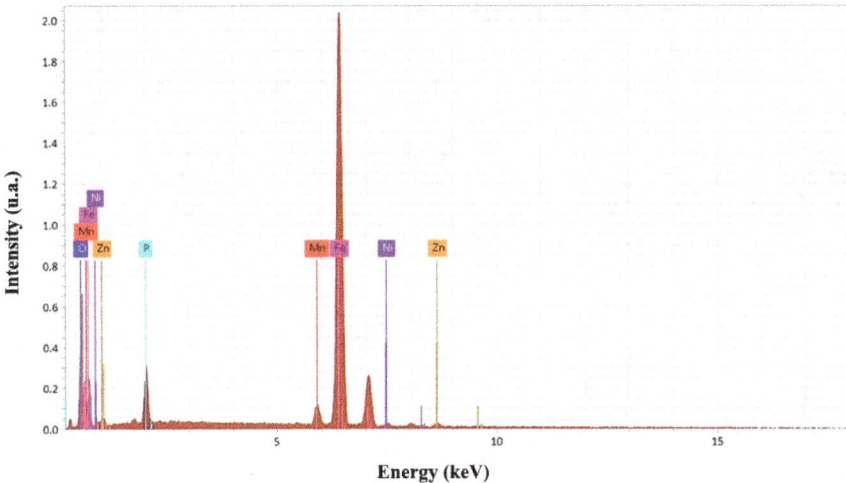

b.

*Figure 3.5.   EDX analysis of the sample phosphated with III/Mn phosphating solution,
(a) elemental chemical distribution, (b) EDX spectrum.*

EDX spectra provide atomic percent values for the constituents, thus confirming the presence of Mg, Zn, Mn, and Ni in the area where the samples exhibited phosphate layers.The presence of iron could be attributed to the formation of crystalline phosphophyllite structures or to the penetration of X-rays into the interface with the substrate, which is rich in Fe. The EDX analysis thus provides unequivocal evidence that the three phosphating solutions used resulted in the formation of thin surface layers on the OB37 steels employed in the reinforced concrete structures.

3.3   X-ray diffraction analysis of chemical compounds after the phosphating process

X-ray diffraction is an analytical technique used for the characterisation of solid, organic or inorganic materials. The diffraction intensities are generated as a result of X-rays interacting with the electrons of the atoms in the lattice.

The selection of powder diffraction or single crystal X-ray diffraction is contingent on the properties of the compounds to be investigated; the information extracted by these two diffractometric techniques is then utilised for the structural characterisation of the crystal lattice [4].

A significant advantage of X-ray diffraction is the rapidity with which it facilitates the acquisition of information; the method is non-destructive and requires only a small sample.The information that can be extracted from X-ray powder diffractograms includes:

- qualitative crystal phase analysis: based on the fact that each crystal phase has a specific diffractogram;

- quantitative phase analysis: if there are several phases in a sample, the number and intensity of the diffraction lines can be used to determine the percentage of each phase identified in the sample under investigation;

- microstructural analysis: determining crystallite sizes, lattice stresses and defect probabilities;

- the degree of crystallinity: is determined by the relationship between the area of diffraction maxima and the total area, which encompasses both the diffraction maxima and the area of the amorphous phase halos;

- determining the crystalline structure of powders.

In the context of single-crystal X-ray diffraction, the crystal structure of the compound under investigation is obtained.

The X-ray diffraction analysis of the phosphate samples was conducted using a diffractometer, which is part of the equipment of the Special Alloys Elaboration Laboratory at the Center for Research and Eco Metallurgical Expertises of the POLITEHNICA University of Bucharest. The results of this analysis are presented in Fig. 3.6, 3.7 and 3.8.

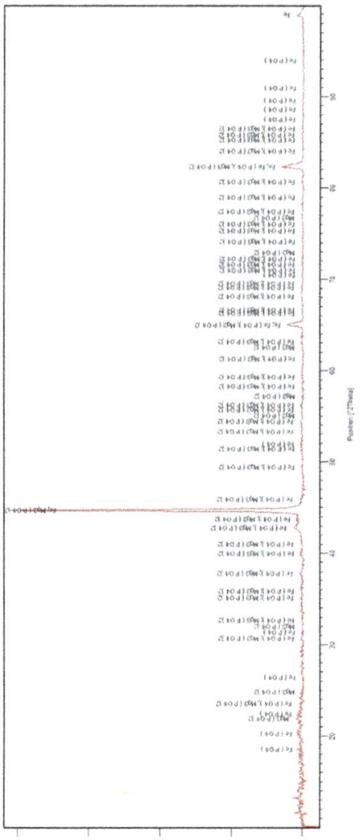

Figure 3.6. Diffractogram of the sample phosphated with I/Mg phosphating solution.

Figure 3.7. Diffractogram of the sample phosphated with II/Zn phosphating solution.

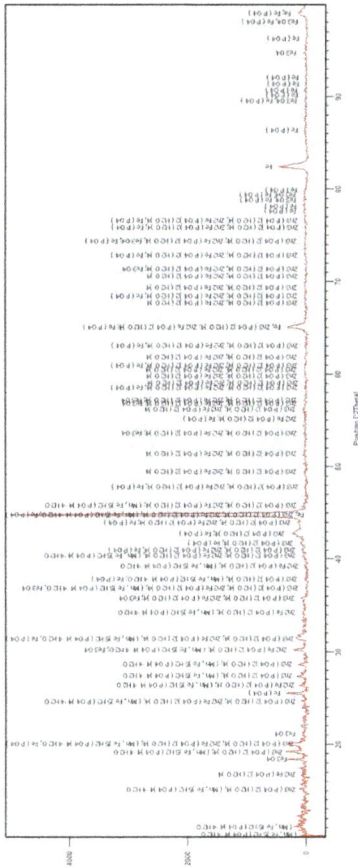

Figure 3.8. Diffractogram of the sample phosphated with III/Mn phosphating solution.

The following compounds were identified in the analysis of the diffractograms of the OB37 steel samples used in reinforced concrete reinforcement: for solution I -

$Mg_2(PO_4)_2$, for solution II - $Zn_3(PO_4)_2x4H_2O$ and $Zn_2Fe(PO_4)_2x4H_2O$, for solution III - $Zn_3(PO_4)_2x4H_2O$, $Zn_2Fe(PO_4)_2x4H_2O$ and $(Mn,Fe)5H2(PO_4)_2x4H_2O$.

3.4 Conclusions

The chemical composition of the layers obtained by immersion in the three phosphating solutions was analysed, as well as that of the compounds formed after phosphating. The deposited layer was structurally analysed by scanning electron microscopy, and EDX and X-ray diffraction analysis were used to reveal the chemical composition of the layers and of the compounds formed after phosphating.

References

[1]   Lazar P., Cimpoesu N., Istrate B., Cazac A.M., Burduhos-Nergis D.-P., Benchea M., Berbecaru A.C., Badarau G., Vasilescu G.D., Popa M. and Bejinariu C. Microstructural and Mechanical Properties Analysis of Phosphate Layers Deposited on Steel Rebars for Civil Constructions. Coatings (Coatings), 2024, Volume 14, Issue 2, Article Number 182. DOI https://doi.org/10.3390/coatings14020182

[2]   Abd El Haleem S.M., Abd El Wanees S., Abd El Aal E.E., Diab A., Environmental factors affecting the corrosion behavior of reinforcing steel II. Role of some anions in the initiation and inhibition of pitting corrosion of steel in Ca(OH)2 solutions, Corros. Sci. 52 (5) (2010) 292–302, http://dx.doi.org/10.1016/j.corsci.2010.01.021

[3]   American Standards Association, American Society of Mechanical Engineers, Society of Automotive Engineers, Surface Texture: Surface Roughness, Waviness, and Lay; American Society of Mechanical Engineers: New York, NY, USA, 2020; ISBN 978-0-7918-7325-0. OCLC 1197629204.

[4]   Information on https://cit.itim-cj.ro/wp-content/uploads/2020/03/07-Difractie-de-raze-X.pdf

CHAPTER 4

Mechanical properties characterisation of deposited coatings

P. Lazar[1], A.-M. Cazac[1]*, C. Bejinariu[1,2]

[1]Faculty of Materials Science and Engineering, Gheorghe Asachi Technical University of Iasi, Romania

[2]Academy of Romanian Scientists, Ilfov 3, 050044 Bucharest, Romania

alin-marian.cazac@academic.tuiasi.ro

Abstract

Chapter four is focused on analysing the mechanical properties of the phosphate coatings. For this purpose, the results of roughness, scratch and micro-hardening tests are analysed. The enhancement of adhesion and surface properties of OB37 steel bars employed in concrete reinforcement represents a pivotal concern for entities engaged in civil construction. This is primarily driven by the loss of material deposition or corrosion that occurs during operation, which can result in catastrophic building collapses and even loss of life. One potential solution to this problem is the application of phosphating, a cost-effective and efficient method of enhancing the quality of the materials - phosphating.

Keywords

Phosphating, Properties Characterisation, Deposited Coatings

4.1 Mechanical properties analysis by roughness tests

The general surface aspects of the samples were obtained by profilometry (Fig. 4.1. a-d). Previous studies have indicated that the crystal size, as well as the type of crystal formed during phosphating, are contingent on the surface roughness of the substrate. This has also been demonstrated to occur after the blasting process, if performed [1], However, in the case of this study, the author aimed to obtain an increase in the surface roughness by phosphating in comparison to the surface roughness of concrete steel concrete, which is commonly applied in practice.

The obvious reason for this is that increasing the surface roughness of the reinforcing elements leads to an improvement in the adhesion of the cement matrix and the formation of a better bond between the concrete steel and the concrete matrix.

The surface profiles of the phosphate samples demonstrate the inhomogeneities of the deposits that are characteristic of this process, yet exhibit a generally homogeneous coverage over the entire surface (Fig. 4.1. b-d). One of the most frequently employed parameters for characterising a surface profile is R_a, a dimensional unit usually given in mm or μm..

R_a is a numerical constant that represents the arithmetic mean of the profile height deviations from the mean line [2,3].

The values of the roughness profile, expressed as R_a, ranged from 0.14 μm (for the sample phosphated with solution I) to 1.37 μm (for the sample phosphated with solution III), all of which are higher than the value of the metallic substrate and may affect the coefficient of friction, microhardness and wear resistance of the tested samples.

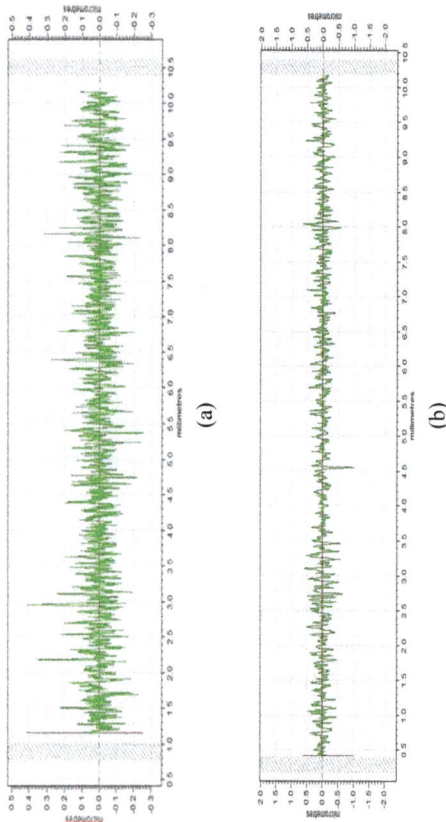

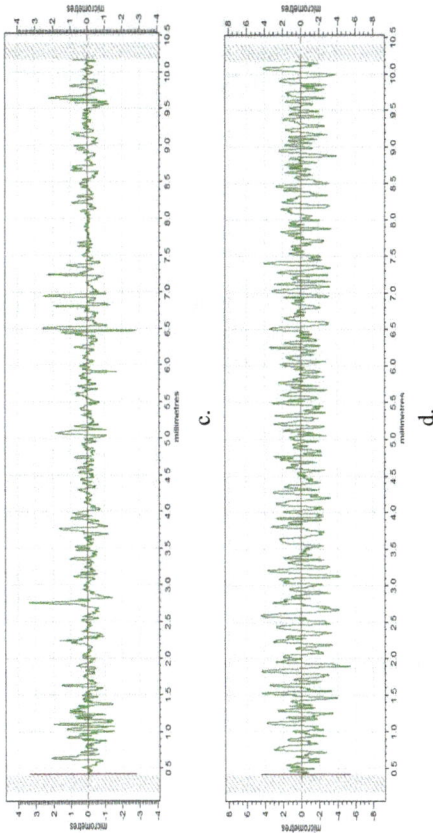

Figure 4.1. Surface profilometry for (a) initial and phosphated metal surface (b) solution I, (c) solution II and (d) solution III.

The highest and lowest average roughness amplitudes in the height direction (Rq) were observed for the samples that were phosphated with solution II (1.67 µm) and solution I (0.17 µm), respectively. It was observed that all R_q values obtained on the phosphated samples were higher than the original sample. The surface characteristics in the height direction, R_{sk} and R_{ku} (asymmetry and kurtosis), were also examined.

Table 4.1. presents the surface roughness profile parameters of the deposited layers and that of the substrate.

Table 4.1.   Surface roughness profile parameters of deposited layers and substrate

Sample	R_a [μm]	R_{sk}	R_q [μm]	R_{ku}
Substrate	0.05	0.54	0.07	4.71
MgCo₃	0.14	−0.14	0.17	0.17
Zn	1.37	−0.01	1.67	0.57
MnZnNi	0.37	1.22	0.55	7.99

As shown in Table 4.1, the values of the asymmetry parameter (R_{sk}) for all samples is directly influenced by the distribution of the phosphate layer on the surface in relation to the mean line. When the phosphate layer is more above the mean line, the value of the asymmetry parameter is negative, and when it is more below the mean line, the value of the asymmetry parameter is positive.

In the present case study, the samples that were subjected to phosphating with solution I and solution II exhibited negative values for the R_{sk} factor. Conversely, the metal substrate and the surface that were phosphated with solution II demonstrated positive values for this parameter. It is noteworthy that surfaces exhibiting asperity removal or areas below the centreline also led to negative values for this parameter. The profile of the sample treated with solution III exhibited a high degree of balance, with the value of the asymmetry parameter approaching zero.

In the context of surface analysis, the study by Herbath (2019) has identified a correlation between the asymmetry parameter and the presence of large peaks or small indentations on the surface profile [4]. From this perspective, the surface obtained by phosphating with solution II may be more suitable for improving the adhesion between the metal insertion bars and the cast concrete matrix in civil engineering.

The kurtosis parameter (R_{ku}), as outlined in Table 4.1, provides additional information regarding the geometry of the investigated surfaces' profiles. Consequently, if the profile falls within the established reference length parameters, it is characterised by a minimal number of peaks and pronounced depths, giving a „platikurtic"- profile with $R_{ku} < 3$; samples phosphated with solutions I and II were found in this situation; furthermore, if the profile has a large number of high peaks and pronounced depths, it is characterised by a „lep-tokurtic" profile with $R_{ku} > 3$.

Initial samples exhibited a parameter R_{ku} greater than 3, and some even showed an R_{ku} greater than 4. These results suggest that the phosphating procedure with solution III results in samples with high R_{ku} values (approximatly 8). This procedure is thus recommended for obtaining a surface with a high adhesion capacity.

4.2 Determination of mechanical characteristics by scratch tests

The objective of the scratch tests was to conduct a preliminary evaluation of the coefficient of friction (COF) of the surface of the investigated samples and to ascertain the adhesion of the outer phosphate layer to the metallic substrate [4,5].

Scratch tests were conducted over a length of 11 mm for all samples and exhibited a differentiated behaviour between the metallic and phosphate layered material, as illustrated in Fig. 4.2. The variation of stress and friction were used to obtain and plot the coefficient of friction.

The variation of the COF of the metallic substrate, which can be considered as a background signal, is added for the samples with phosphated layers and the behaviour of each of the layers obtained as a function of the solution used (I, II or III). With the exception of a pronounced decline in the COF of the phosphated sample when utilising solution III, as illustrated in the variation plot of Fig. 4.2., the samples exhibited similar behaviour. The decrease in the value of the COF in this case, which is very close to the values obtained on the metallic substrate, may be due to the partial disruption of the phosphate layer in that area.

The elevated variability in the values of the COF on the phosphated samples is evidently attributable to the phosphate layer, which is partially diffused within the metallic substrate and partially deposited on the surface. This is attributable to the ceramic nature of these layers, which is more brittle in comparison to the metallic substrate.

As illustrated in Fig. 4.2, for the initial 2-3 mm of the scratch, the samples that were phosphated with solutions I and III exhibited a slightly higher COF than the metallic substrate, to distinguish them from the sample that was phosphated with solution I, which exhibited a higher coefficient of friction over this distance.

Following an increase in the scratch stress and potential fracture of the phosphate layer on the surface of OB37 steel (used in concrete reinforcement), the three samples exhibited comparable behaviour, although with slightly higher values for the sample phosphated with solution II. Despite the penetration of the outer phosphate layer by the test indenter, the metallic substrate exhibited an upper layer through which the phosphate layer diffused. This layer exhibited a higher COF compared to the substrate in its original state.

It is important to note that both static and kinetic friction have different values when referring to the COF. Static friction is defined as the resistance encountered when two objects are at rest and attempt to move [6-8]. The friction force is thus resistive, preventing objects from slipping until a critical force is applied. The force required to initiate motion is known as static frictional force.

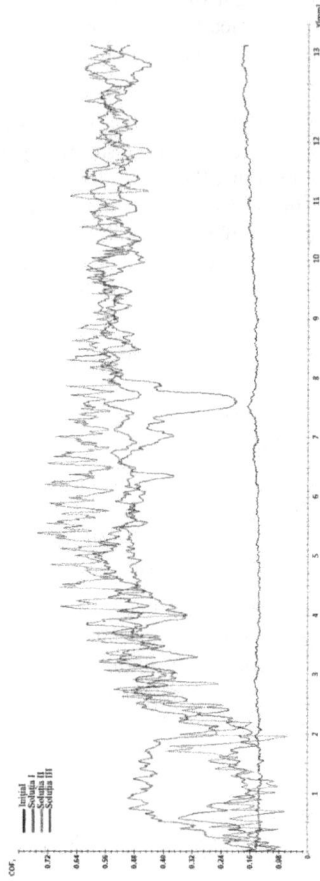

Figure 4.2. Variation of COF with distance for: initial metal surface and surfaces phosphated with phosphating solutions I, II and III.

Kinetic friction, otherwise referred to as sliding friction, is defined as the force that opposes the motion of an object. The kinetic frictional force is typically smaller than the static frictional force. To illustrate this principle, consider the motion of a brick sliding on a mass of wood. In this scenario, the coefficient of kinetic friction is 0.5, indicating that a force equal to at least half the weight of the brick is required to maintain a constant speed of movement [9-11].

Conversely, the coefficient of static friction for the aforementioned case is approximately 0.6. In both scenarios, the friction force is oriented in the opposite direction to the object's motion. It is noteworthy that materials exhibiting a COF lower than 0.1 are classified as lubricating materials [12].

The mean values of the results obtained in the micro-scratch test of the phosphate layers with the three solutions and the non-phosphate sample are given in Table 4.2.

Table 4.2.   Results of the micro-scratch test of deposited layers (mean values)

Phosphating solution	F_x [N]	Cof
Initial	0.845	0.149
Solution I/Mg	2.862	0.491
Solution II/Zn	2.588	0.436
Solution III/Mn	2.937	0.493

As illustrated in Fig. 4.3, the absence of phosphating is evident in the scratch marks on the sample surfaces. In contrast, the presence of phosphating, along with the utilisation of various experimental solutions, is shown in Fig. 4.3, (b-d).

The SEM images of the traces (Fig. 4.3.(a-d)) provide a visual confirmation of the variations in the COF obtained by the scratch test. The distribution of elements on the surface is also illustrated in Fig. 4.3.(a-d). Structurally, the traces exhibit a high degree of similarity, with an approximate thickness of 75-80 μm and an absence of breaks. However, one exception was observed, namely the sample that was phosphated with solution II, as shown in Fig. 4.2. (c) and 4.3. (c), where the phosphate layer could be seen after scratching.

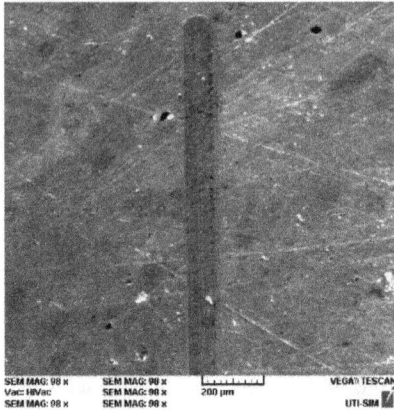

a.

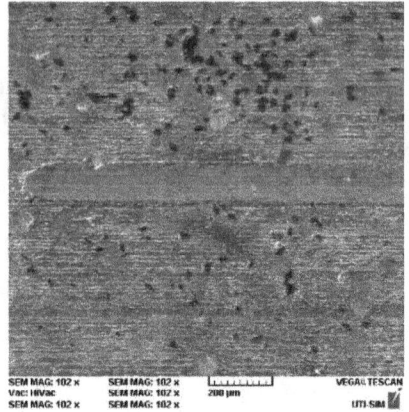

b.

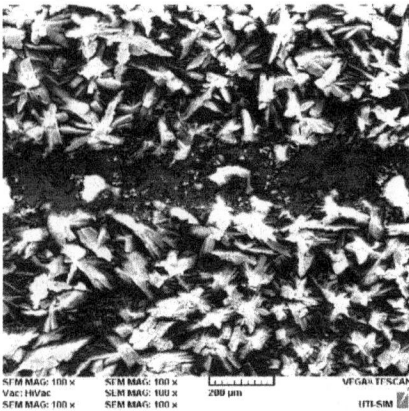

c.

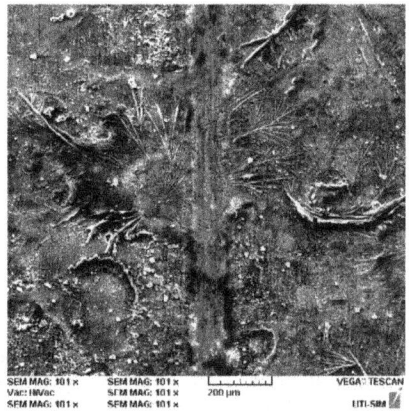

d.

Figure 4.3. SEM images of the scratch: (a) non-phosphate surface and for the phosphate surfaces (b) with solution I, (c) with solution II and (d) with solution III.

In all cases, when the surface was scratched, no evidence of large cracks in the phosphate layer was observed; Fig. 4.4. It was found that all the phosphate layers were pierced and removed by the scratch test.

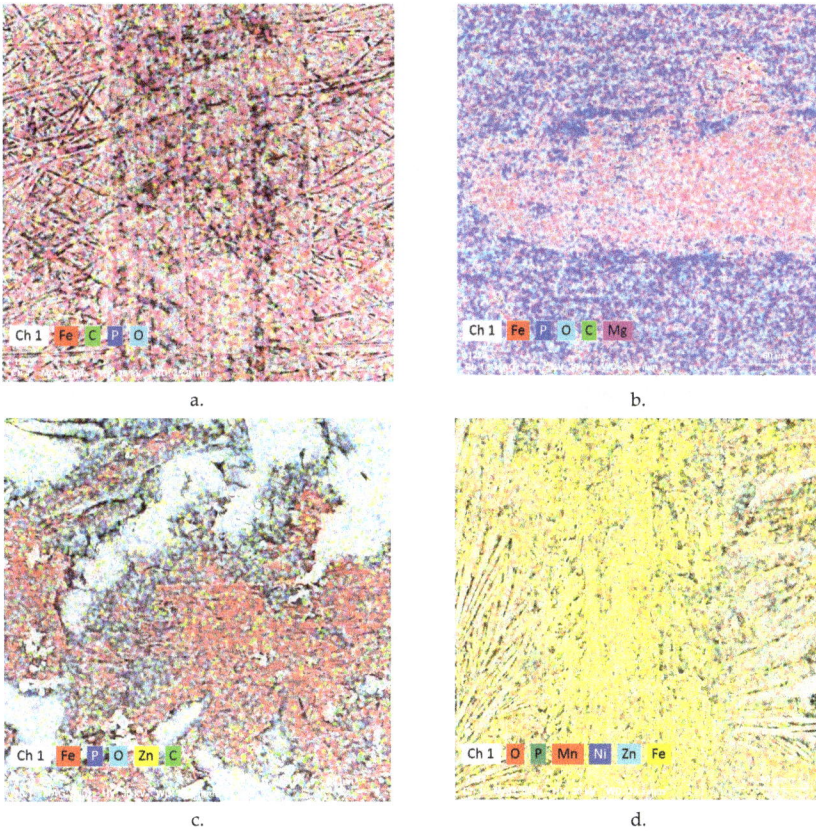

Figure 4.4. EDS images of elemental distributions on the scratched areas: (a) non-phosphate surface and for the phosphate surfaces (b) with solution I, (c) with solution II and (d) with solution III.

The elemental distributions provide confirmation of elemental diffusion from the phosphating solution to the top of the metal substrate during deposition, the same elements being present at the scratch marks where the outer phosphate layer has been mechanically removed by indentation. Samples that were phosphated with solution II, as illustrated in Fig. 4.4.(c), exhibit areas where the phosphate layer exhibited resistance to the external scratch stress and was not completely removed by the indenter. This phenomenon persists despite the presence of reduced microhardness at the surface, yet the deposited phosphate layer demonstrates superior elasticity.

4.3   Mechanical characteristics determination by microindentation tests

The microindentation test was conducted utilising the CETR UMT-2 tribometer (Ettlingen, Germany), employing the Rockwell method. The initial preload stage, performed with a force of 1% of the maximum test force, was achieved within a time period between 10 s and 30 s. The maximum loading force used in this study was 9 N. A total of three tests were performed on the unphosphated OB37 steel specimen, the values obtained are shown in Table 4.3. The mean hardness value obtained was 1.59 GPa, the mean modulus of elasticity was 148.46 GPa, and the mean average contact stiffness value was 5.17 μm.

Table 4.3.   Microindentation test results

	Nr. crt.	Young indentation modulus [GPa]	Hardness [GPa]	Maximum load [N]	Maximum displacement [μm]	Rigidity [N/μm]	Contact area [μm²]
Initial	Point 1	92.55	1.63	8.99	5.31	7.76	5517.57
	Point 2	212.19	1.54	9.00	5.12	16.56	5856.25
	Point 3	140.64	1.62	8.99	5.07	11.33	5559.55
	Average	148.46	1.59	8.99	5.17	11.88	5644.45
Solution I/Mg	Point 1	150.64	1.54	9.02	5.27	12.36	5866.21
	Point 2	126.15	1.63	9.02	5.12	10.27	5539.07
	Point 3	187.87	1.46	9.03	5.41	15.34	6166.61
	Average	154.89	1.54	9.02	5.27	12.65	5857.29
Solution II/Zn	Point 1	86.29	0.98	9.02	8.17	9.33	9181.95
	Point 2	100.39	1.13	9.00	7.12	10.02	7974.12
	Point 3	167.28	1.38	9.00	5.72	14.26	6507.06
	Average	117.98	1.16	9.01	7.01	11.21	7887.71
Solution III/Mn	Point 1	221.41	1.42	9.02	5.52	17.89	6377.38
	Point 2	118.57	1.69	8.99	5.00	9.53	5335.96
	Point 3	103.51	1.42	9.02	5.84	9.18	6337.26
	Average	147.83	1.51	9.01	5.45	12.20	6016.86

As illustrated in Fig. 4.5, the specific sample with the values closest to the mean values demonstrates a residual deformation of approximately 4.72 μm. This system exhibits plastic behaviour without the presence of macrocracks.

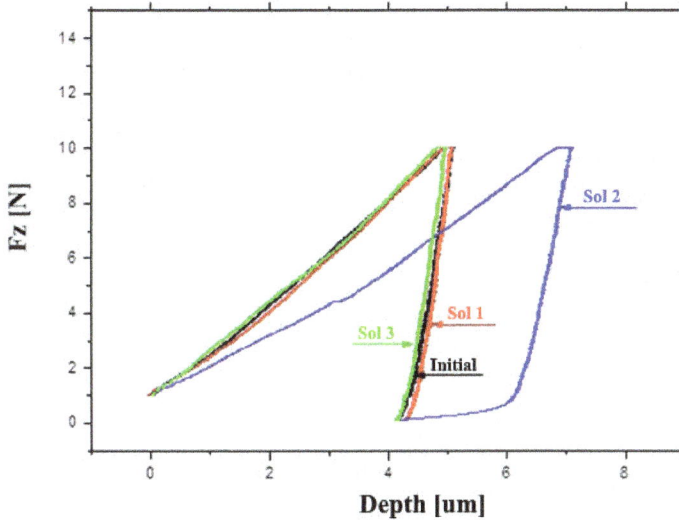

Figure 4.5.　Indentation plots of the initial metal surface and with phosphating solutions I/Mg, II/Zn and III/Mn.

As demonstrated in Fig.e 4.5, the values of elastic modulus and hardness are in accordance with the graph, where it can be seen that the initial sample had the lowest maximum deformation and the solution III samples had the highest values. The initial sample exhibited an average hardness of 1.595 GPa, accompanied by an elastic modulus of 148.46 GPa. In contrast, the solution II samples demonstrated a hardness of 1.165 GPa, with a Young's modulus of 117.98 GPa, both of which were lower than those observed in the solution I and III samples.

The depths of indentations, Fig. 4.5, are mainly in the phosphate coating layer (typically 20-30 μm), with a greater depth for samples phosphated with the second solution (solution II). It was observed that both coatings obtained after phosphating with solution I and III exhibited values of Young's modulus of indentation that were close to those of the metallic CS material, which is considered to be a significant advantage for the mechanical properties of the deposited material. The hardness values exhibited a homogeneous layer in all cases, with a lower hardness value observed for the sample

65

phosphated with solution II. With regard to contact stiffness, the sample phosphated with the second solution exhibited the most favourable outcomes, demonstrating a lower value in comparison to the metallic substrate [4]. However, it should be noted that all samples exhibited comparable values, as illustrated in Table 4.5.

4.4   Conclusions

The enhancement of adhesion and surface properties of OB37 steel bars employed in concrete reinforcement represents a pivotal concern for entities engaged in civil construction. This is primarily driven by the loss of material deposition or corrosion that occurs during operation, which can result in catastrophic building collapses and even loss of life. One potential solution to this problem is the application of phosphating, a cost-effective and efficient method of enhancing the quality of the materials - phosphating.

Three distinct solutions were employed to obtain phosphate coatings on a construction steel substrate at an acceptable cost and time. The experimental results led to the following main conclusions:

• The samples that were phosphated with phosphate solution I/Mg and phosphate solution III/Mn showed negative values for the Rsk factor, while the metal substrate and the surface phosphated with phosphate solution II/Zn showed a positive value. The profile of the sample phosphated with phosphate solution III/Mn showed a very balanced profile and the value of the asymmetry parameter was very close to zero;

• The OB37 control sample had the lowest maximum strain and the III/Mn phosphated sample had the highest. The initial sample exhibited an average hardness of 1.595 GPa, with a Young's modulus of 148.46 GPa. In contrast, the samples phosphated with phosphate solution II/Zn demonstrated a hardness of 1.165 GPa, with a Young's modulus of 117.98 GPa, both of which were lower than those observed for the phosphate samples phosphated with phosphate solutions I/Mg and III/Mn;

• The coefficient of friction obtained for the phosphated samples shows values three times higher (0.49 for the phosphated samples with phosphate solutions I/Mg and III/Zn, respectively 0.44 for the sample with phosphate solution II/Zn) than the average value obtained on the metallic substrate (0.15);

• The phosphated surface that was subjected to treatment with phosphating solution II/Zn exhibited areas where the phosphate layer had not completely removed from the indenter due to external scratching.

A thorough analysis of the outcomes derived from the experimentation with the OB37 steel samples that were subjected to phosphating with the solutions under consideration reveals that the most optimal outcomes for the evaluated tests were attained on the samples that underwent phosphating with the solution II/Zn. It can thus be concluded that a solution comprising 9 g of Zn, 2 l, can be utilised to obtain steel bars for application in civil construction, thereby enhancing their surface properties.

References

[1]   Bejinariu C, Lazar P, Sandu AV, Cazac AM, Sandu IG, Corbu O, Enhancing
        Properties of Reinforcing Steel by Chemical Phosphatation. The ICAMET 2014,
        3th International Conference Proceedings – Advanced Materials Engineering &
        Technology, Ho Chi Minh City, Vietnam, December 4-5, 2014, Applied
        Mechanics and Materials Vols. 754-755 (2015) pp 310-314, (2015) Trans Tech
        Publications, Switzerland, https://doi.org/10.4028/www.scientific.net/AMM.754-
        755.310, ISBN: 978-3-03835-434-5

[2]   Abd El Haleem S.M., Abd El Wanees S., Abd El Aal E.E., Diab A.,
        Environmental factors affecting the corrosion behavior of reinforcing steel II. Role
        of some anions in the initiation and inhibition of pitting corrosion of steel in
        Ca(OH)2 solutions, Corros. Sci. 52 (5) (2010) 292–302, https://doi.org/10.1016/
        j.corsci.2010.01.021

[3]   Duffó, G.; Reinoso, M.; Ramos, C.; Farina, S. Characterization of steel rebars
        embedded in a 70-year old concrete structure. Cem. Concr. Res. 2012, 42, 111–
        117. https://doi.org/10.1016/j.cemconres.2011.08.003

[4]   Lazar P., Cimpoesu N., Istrate B., Cazac A.M., Burduhos-Nergis D.-P., Benchea
        M., Berbecaru A.C., Badarau G., Vasilescu G.D., Popa M. and Bejinariu C.
        Microstructural and Mechanical Properties Analysis of Phosphate Layers
        Deposited on Steel Rebars for Civil Constructions. Coatings (Coatings), 2024,
        Volume 14, Issue 2, Article Number 182.
        https://doi.org/10.3390/coatings14020182

[5]   Bulbuc, V.; Paleu, V.; Pricop, B.; Popa, M.; Cârlescu, V.; Cimpoesu, N.;
        Bujoreanu, L.G. Variation of wear resistance of T105Mn120castings, used for
        railway safety systems, as an effect of dynamic loading under extreme conditions.
        JMEP 2021,30, 7128–7137

[6]   Burduhos-Nergis D.-P., Sandu A.V., Burduhos-Nergis D.-D., Vizureanu P.,
        Bejinariu C., Phosphate Conversion Coating – A Short Review, Arch. Metall.
        Mater. 68 (2023), 3, 1029-1034, https://doi.org/10.24425/amm.2023.145471

[7]   Burduhos-Nergis D.P., Vasilescu G.D., Burduhos-Nergis D.D., Cimpoesu R.,
        Bejinariu C., Phosphate coatings: EIS and SEM applied to evaluate the corrosion
        behavior of steel in fire extinguishing solution, 2021, Applied Sciences 11 (17),
        7802.

[8]   Burduhos-Nergis D.P., Vizureanu P., Sandu A.V., Bejinariu C. Evaluation of the
        corrosion resistance of phosphate coatings deposited on the surface of the carbon
        steel used for carabiners manufacturing, 2020, Applied Sciences 10 (8), 2753.

[9]   Samardžija M.; Alar V.; Aljinović F.; Kapor F., Influence of phosphate layer on
        adhesion properties between steel surface and organic coating. Rud. Geološko
        Naft. Zb. 2022, 37, 11–17. https://doi.org/10.17794/rgn.2022.1.2.55

[10] Hafiz M.H., Kashan J.S., Kani A.S., Effect of Zinc Phosphating on Corrosion Control for Carbon Steel Sheets, 2008, Eng. & Technology, 26(5), 501-511.

[11] Ernens D., de Rooij M.B., Pasaribu H.R., van Riet E.J., van Haaften W.M., Schipper D.J., Mechanical characterization and single asperity scratch behaviour of dry zinc and manganese phosphate coatings, Tribology International 118 (2018) 474–483, https://doi.org/10.1016/j.triboint.2017.04.034

[12] Królikowski A., Kuziak J., Impedance study on calcium nitrite as a penetrating corrosion inhibitor for steel in concrete, Electrochim. Acta 56 (23) (2011) 7845–7853, http://dx.doi.org/10.1016/j.electacta.2011.01.069

Materials Research Foundations **183** (2025)   https://doi.org/10.21741/9781644903810

CHAPTER 5

Investigation of the corrosion behaviour of phosphate layers deposited on OB37 steel

P. Lazar[1], A.-M. Cazac[1]*, C. Bejinariu[1,2]

[1]Faculty of Materials Science and Engineering, Gheorghe Asachi Technical University of Iasi, Romania

[2]Academy of Romanian Scientists, Ilfov 3, 050044 Bucharest, Romania

alin-marian.cazac@academic.tuiasi.ro

Abstract

Chapter five examines the corrosion behaviour of phosphate coatings deposited on OB37 steel used in concrete reinforcement. It is widely acknowledged that for reinforced concrete structures, steel corrosion due to general exposure to oxygen and moisture poses the most significant threat to their stability and durability. In the present case study, the corrosion behaviour of OB37 steel was investigated by subjecting the steel to two distinct corrosion environments: one composed of fresh water extracted from the Bahlui River and another comprising sea water drawn from the Aegean Sea.

Keywords

Phosphating, Corrosion Behaviour, Phosphate Layers

5.1 Corrosion behavior evaluation

The OrigaFlex01A electrochemical system, in conjunction with a computer and OrigaMaster 5 software for data acquisition and processing, was utilised for the measurements. The following electrochemical methods were used to analyse the corrosion behaviour: open circuit potential (OCP), linear voltammetry (LV), and cyclic voltammetry (CV). The use of potentiostats, in conjunction with the operating system, facilitates the evaluation of several parameters, including: the polarization resistance, the corrosion potential over extended periods, the torque potential, and the detection of both pitting and generalised corrosion.

The anodic polarization curves were obtained at low scanning rates of the electrode potential so that the electrochemical metal/solution system is as close to equilibrium as possible and the curves allow the most accurate evaluation of polarisation resistance, instantaneous corrosion current and instantaneous corrosion rate. For this purpose, the potential scanning rate used was: $dE/dt = 0.5$ mV/s. The area of exposure to the corrosion medium was S = 0.503 cm^2. Linear voltammetry was performed in a range of ±250mV with respect to the OCP and cyclic voltammetry in a range of -200---+700mV with a scanning rate of 5 mV/s.

As illustrated in the following figure, the three-electrode cell and working electrode were used for measurements, Fig. 5.1.

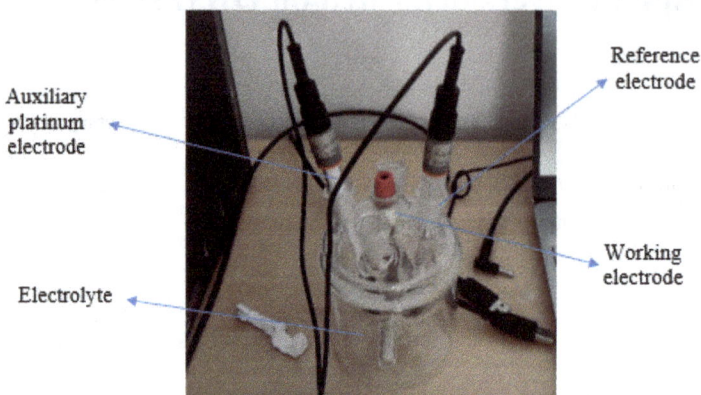

Figure 5.1.   The three-electrode cell used.

The auxiliary electrode, which is used to polarize the working electrode, is composed of platinum. The reference electrode, against which the potential of the working electrode is measured, is a saturated calomel electrode (SCE), which demonstrates excellent stability over time.

To study the corrosion behaviour, tests were performed in two electrolyte solutions: water from the Bahlui River and water from the Aegean Sea, both for the initial OB37 steel sample and for the samples coated with the three phosphating solutions [1, 2].

5.1.1.   Corrosion behavior of unphosphated control samples

The polarisation diagrams for OB37 in seawater and Bahlui water electrolyte solution are shown in Fig. 5.2. In both instances, it is noticeable that the anodic reaction, the oxidation of Fe in this case reaction (5.1), is predominant compared to the cathodic reaction reaction (5.2). It is important to note that in these aqueous environments, corrosion occurs in the presence of oxygen.

The corrosion of Fe in aqueous media is initiated by the following anodic reaction:

$$Fe \rightarrow Fe^{+2} + 2e^-$$ (5.1)

and the cathodic reaction:

$H_2O + 1/2O_2 + 2e^- \rightarrow 2OH^-$ (5.2)

The initial layer located near the Fe surface consists of Fe and O. This layer is initially formed after the Fe⁻ ions and OH⁻ reactions (5.1) and (5.2) proceed as follows:

$Fe^{+2} + 2OH^- \rightarrow Fe(OH)_2$ (5.3)

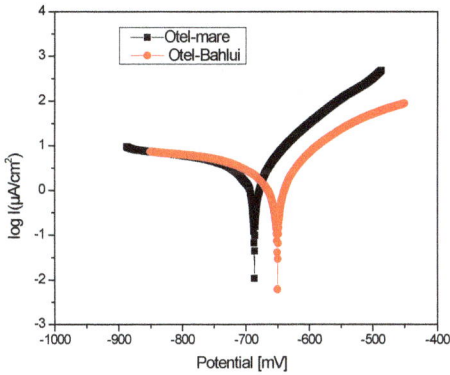

a)

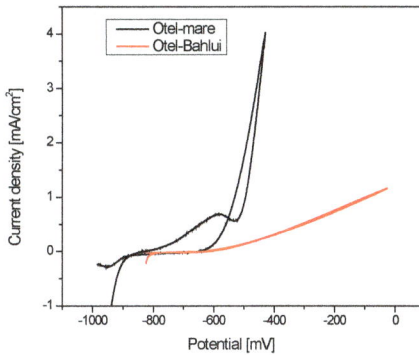

b)

Figure 5.2.   Polarization curves a) Tafel diagram; b) cyclic diagram of steel OB37.

Further oxidation of iron to Fe^{+3} occurs according to the reaction:

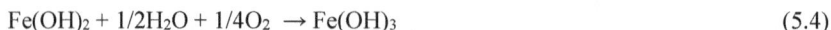

$$Fe(OH)_2 + 1/2H_2O + 1/4O_2 \rightarrow Fe(OH)_3 \qquad (5.4)$$

The reactions (5.1) ÷ (5.4) result in the formation of hydroxides due to corrosion in aqueous solutions.

An analysis of the cyclic diagrams reveals a discrepancy in the behavior of sample OB37 in the two solutions. See Fig. 5.2.(b). The diagram in the Bahlui water indicates a general or generalized corrosion process over the entire surface of the alloy, with no indications of passivation phenomena. In this instance, the current density remains constant and is significantly lower than in seawater. In contrast, the cyclic diagram recorded in the case of seawater exhibits a pitting-corrosion character, with a substantial increase in current density during the oxidation processes of the metal. At higher potentials, in parallel with metal oxidation, oxygen evolution may also occur. This phenomenon may be attributed to the elevated presence of chloride ions in seawater. The autocatalytic reaction between chloride ions (Cl^-) and iron ions ($Fe(II)$) leads to the formation of pitting corrosion [3]:

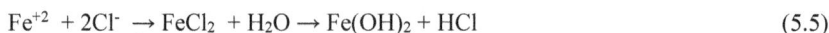

$$Fe^{+2} + 2Cl^- \rightarrow FeCl_2 + H_2O \rightarrow Fe(OH)_2 + HCl \qquad (5.5)$$

5.1.2   *Corrosion behavior of samples phosphated with solution I/Mg*

The potentiodynamic polarization curves for samples phosphated with solution I are shown in Figure 5.3. In this case, the phosphating layer is enriched with Mg ions. A similar behavior is also observed for the linear polarization of the samples in the two solutions, with the anodic reaction predominating [4,5]. The corrosion of Mg in an aqueous solution transpires through a series of three reactions, which are outlined below:

First stage - the anodic reaction:

$$Mg \rightarrow Mg^{2+} + 2e^- \qquad (5.6)$$

Second stage - the cathodic reaction:

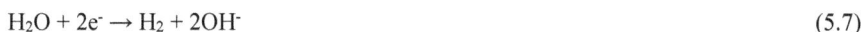

$$H_2O + 2e^- \rightarrow H_2 + 2OH^- \qquad (5.7)$$

And formation of corrosion products:

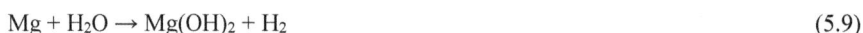

$$Mg^{2+} + 2OH^- \rightarrow Mg(OH)_2 \qquad (5.8)$$

$$Mg + H_2O \rightarrow Mg(OH)_2 + H_2 \qquad (5.9)$$

In the case of cyclic polarization (Fig. 5.3(b)), the behaviour is similar to that of sample OB37, possibly due to the high reactivity of the Mg layer, which resulted in the Mg layer penetrating and the electrolyte reaching the substrate [3].

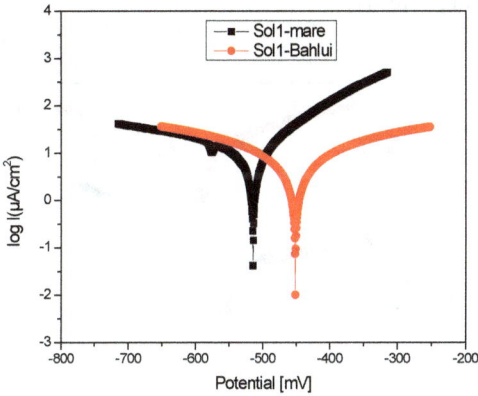

a)

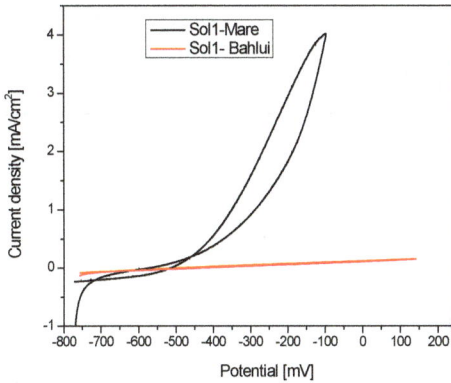

b)

Figure 5.3.   Polarization curves a) Tafel diagram; b) cyclic diagram for I/Mg solution.

5.1.3   Corrosion behavior of samples phosphated with solution II/Zn

Fig. 5.4. shows the linear and cyclic polarisation plots for samples phosphated with solution II in Bahlui and seawater. In this instance, there is no significant disparity between the anodic and cathodic slopes, indicating that the oxidation-metal corroding process is occurring at a substantially reduced rate [6]. This phenomenon can be attributed to the presence of Zn in the phosphate layer, which, when dissolved in water, forms ZnO, a passivating oxide [7].

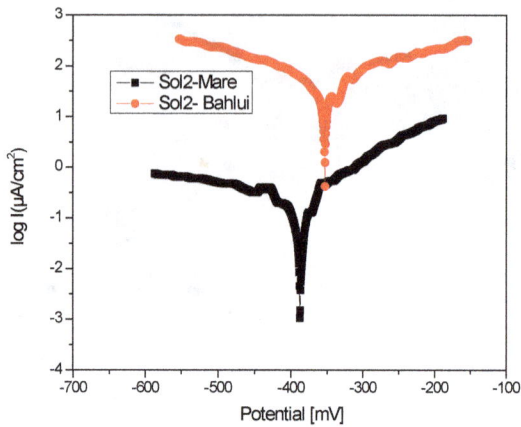

a)

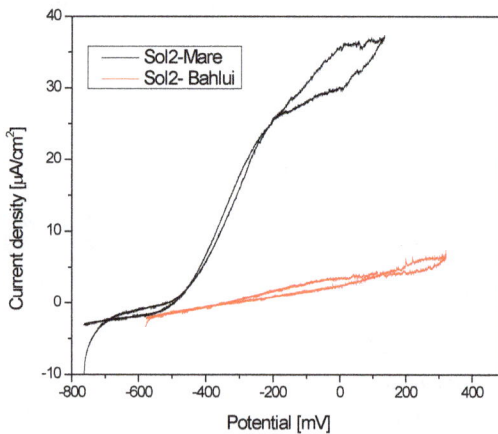

b)

Figure 5.4.   Polarization curves a) Tafel diagram; b) cyclic diagram for solution II/Zn.

The following redox reactions occur during Zn corrosion with an aqueous medium:

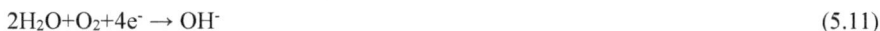

$$Zn \rightarrow Zn^{+2} + 2e^- \qquad\qquad (5.10)$$

$$2H_2O + O_2 + 4e^- \rightarrow OH^- \qquad\qquad (5.11)$$

And the zinc oxide formation reactions:

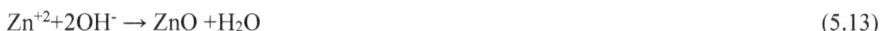

$$Zn^{+2} + 2OH^- \rightarrow Zn(OH)_2 \qquad\qquad (5.12)$$

$$Zn^{+2} + 2OH^- \rightarrow ZnO + H_2O \qquad\qquad (5.13)$$

Fig. 5.4.(b) shows the cyclic polarisation curves recorded in both corrosion environments. The plots are indicative of a pitting corrosion alloy, with a higher current density value in seawater due to Cl ions. It is noteworthy that beyond a certain potential, defined as the pitting potential (approximately -380mV in this instance), the metal exhibits signs of corrosion

5.1.4 *Corrosion behavior of samples phosphated with solution III/Mn*

For the samples that were phosphated with solution III, the Tafel and cyclic plots are presented in Fig. 5.5. As previously observed, there is a negligible difference between the anodic and cathodic slopes, with the layer corroding process occurring at a significantly slower rate than in the case of OB37. A clear difference in the corrosion behaviour in both environments can be seen from the cyclic polarisation curves, with the samples showing a generalised corrosion character. The diagram indicates a corrosion process over the entire surface. The return curve (cathodic branch of the polarization curve) exactly overlaps the cathodic branch, indicating that no other processes (passivation, transpassivation) are occurring in this area. This phenomenon may be attributed to the compositional complexity of the Ni, Fe, and Mn phosphate layer.

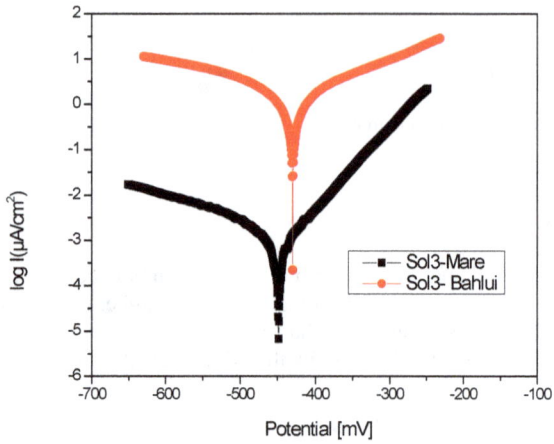

a)

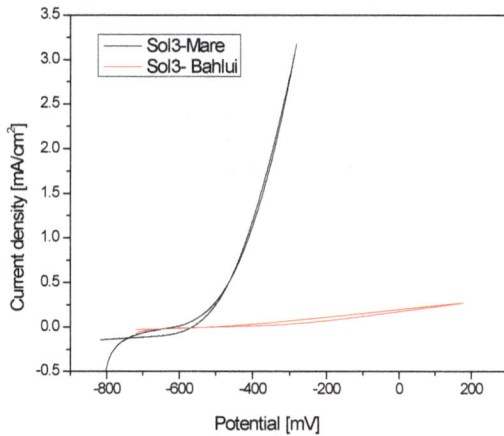

b)

Figure 5.5.   Polarization curves a) Tafel diagram; b) cyclic diagram for solution III/Mn.

5.1.5   Corrosion process parameters synthesis

In Table 5.1. the corrosion process parameters from the table diagrams are shown for all phosphate samples and OB37.

The samples that exhibited the best corrosion resistance were those phosphated with the Zn solution, with the lowest corrosion rate, V_{corr}, in both electrolyte media, **0.258** μm /y for Bahlui water and **3.060** μm /y for seawater. These values are substantially lower than the corrosion rates of steel, which are approximately 45 μm /y.

In relation to the corrosion potential, E (I=0), which is indicative of the corrosion tendency of the alloy when immersed in an electrolytic medium, the samples that were phosphated with the II/Zn solution showed the best behaviour. The lowest corrosion potential value recorded is -353 mV for the II/Zn solution in Bahlui water, compared to -650.6 mV for OB37 steel in the same electrolyte. The corrosion current values determined at the corrosion potential of the alloy using the linear polarization curve represent the corrosion current that occurs at the metal/corrosive medium interface when the metal is immersed in the solution. It is important to note that this corrosion current is inversely proportional to the value of the polarization resistance [4]. The lowest recorded corrosion current was observed in samples that underwent phosphating with the II/Zn solution. Conversely, the I/Mg solution, characterised by its high reactivity, exhibited the most significant corrosion current and oxidation reaction

Table 5.1.   Instantaneous corrosion process parameters versus electrolyte

Parameter	OB37 steel		Sol. I - MgCO$_3$		Sol. II - Zn		Sol. III - NiFeMn	
	Bahlui	Aegean Sea	Bahlui	Aegean Sea	Bahlui	Aegean Sea	Bahlui	Aegean Sea
E(I=0), mV	-650.6	-686.6	-451.1	-514.3	-353	-386.8	-448.0	-429.9
I_{corr}, μA	3.937	3.762	10.393	23.221	0.028	0.187	2.6	3.059
Rp kohm/cm^2	13.25	7.78	5.52	2.64	291	154.18	20.33	18.54
v_{corr}, m/y	45.352	43.341	226.35	421.47	0.258	3.060	34.89	40.82
β_a, mV/dec	129	92	365	144	180	97	67	200
$-\beta_c$, mV/dec	737	543	363	770	102	268	244	350

5.2   Morphology and structure investigation of phosphate layers after corrosion

A scanning electron microscope equipped with an EDAX X-ray detector was utilised to evaluate the morphological surface modification of the samples following corrosion. For the purpose of comparison, analysis was performed on two samples, OB37 steel corroded in seawater and the sample phosphated with solution II in the same electrolyte.

5.2.1 *Morphology and structure analysis of control samples after corrosion*

The surface condition of sample OB37 was examined at various image magnification powers ((a) 100x, (b) 500x şi (c) 1000x) following a corrosion resistance test in Aegean Sea water. The results are presented in Fig. 5.6, which was obtained using SEM microscopy.

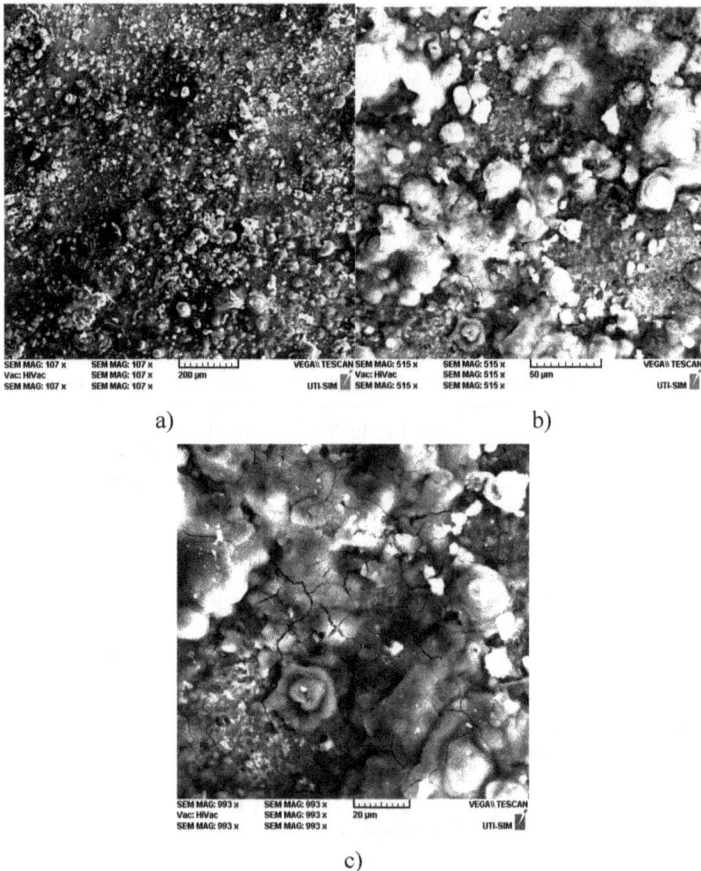

a)   b)

c)

Figure 5.6.   SEM images of OB37 at different image magnification powers after the corrosion process in Aegean Sea water - 100x, b- 500x and c- 1000x..

The corrosion test showed that the material was severely damaged by the deposition of iron oxide and iron hydroxide corrosion compounds on its surface. Furthermore, Fig. 5.7 indicates the presence of chlorine and oxygen on the surface of OB37 steel.

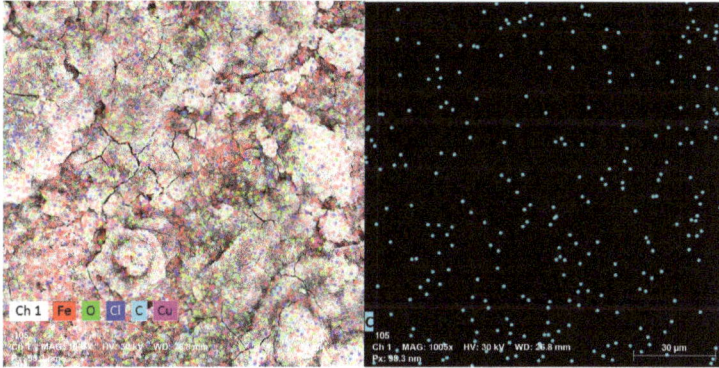

a)   b)

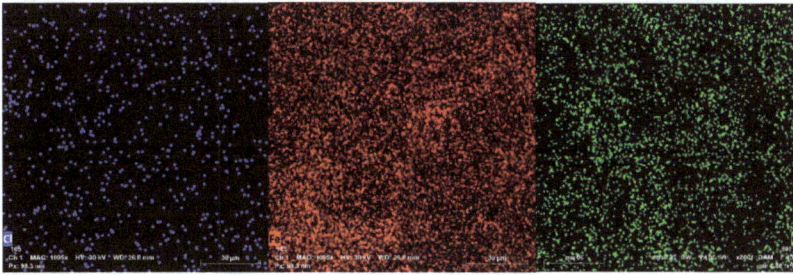

-c-   -d-   -e-

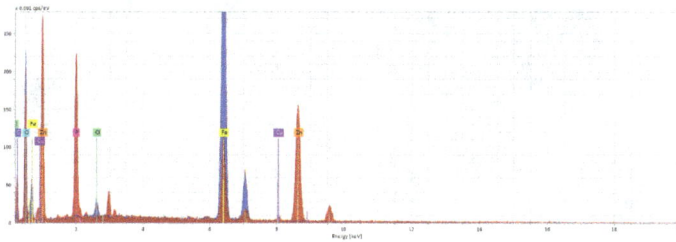

f)

*Figure 5.7.   Main elements distribution on the OB37 surface after the corrosion test a-
element distribution; b- carbon distribution; c- chlorine distribution; d- iron distribution;
e- oxygen distribution; f- energy spectrum.*

Fig. 5.7(f) illustrates the energy spectrum and Table 5.2 provides the relevant chemical composition data. These show high percentages of O and chlorine, and a decrease in iron mass due to the oxidation process

Table 5.2.   Chemical composition of OB37 steel surface after corrosion in Aegean Sea electrolyte solution.

Element	At. No.	Mass Norm. [%]	Atom [%]
Fe	26	57.778	27.26
O	8	33.244	54.85
C	6	7.7081	16.94
Cl	17	1.2623	0.939
Total		100	100

5.2.2   Morphology and structure analysis of samples phosphated with solution II/Zn after corrosion

The SEM microscopy of the sample phosphated with II/Zn solution, which showed the best corrosion resistance at different magnifications after the corrosion process in Aegean Sea water, is shown in Fig. 5.8. It is noted that the coating did not suffer any major damage after the corrosion test, no corrosion marks or exfoliation are observed on the surface. The formation of micron-sized compounds on the surface during the corrosion process can also be observed in this case.

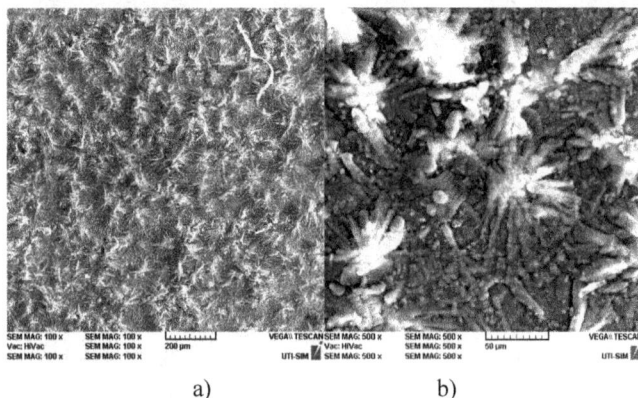

a)   b)

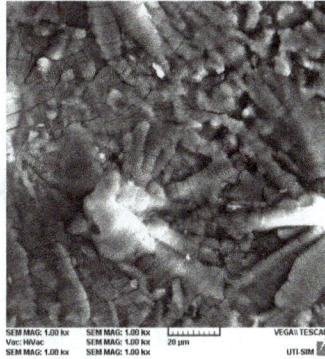

c)

*Figure 5.8.   SEM images of the sample phosphated with solution 2 at different image
magnification powers after the corrosion process in Aegean Sea water: a - 100x, b - 500x
and c - 1000x.*

The elemental distribution in Fig. 5.9. also shows the presence of chlorine, oxygen and
phosphorus on the surface of the layer.

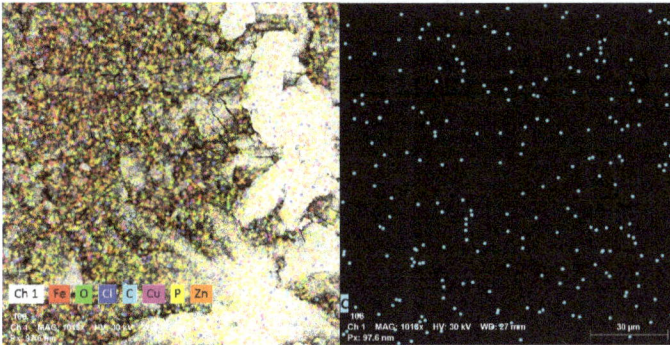

a)   b)

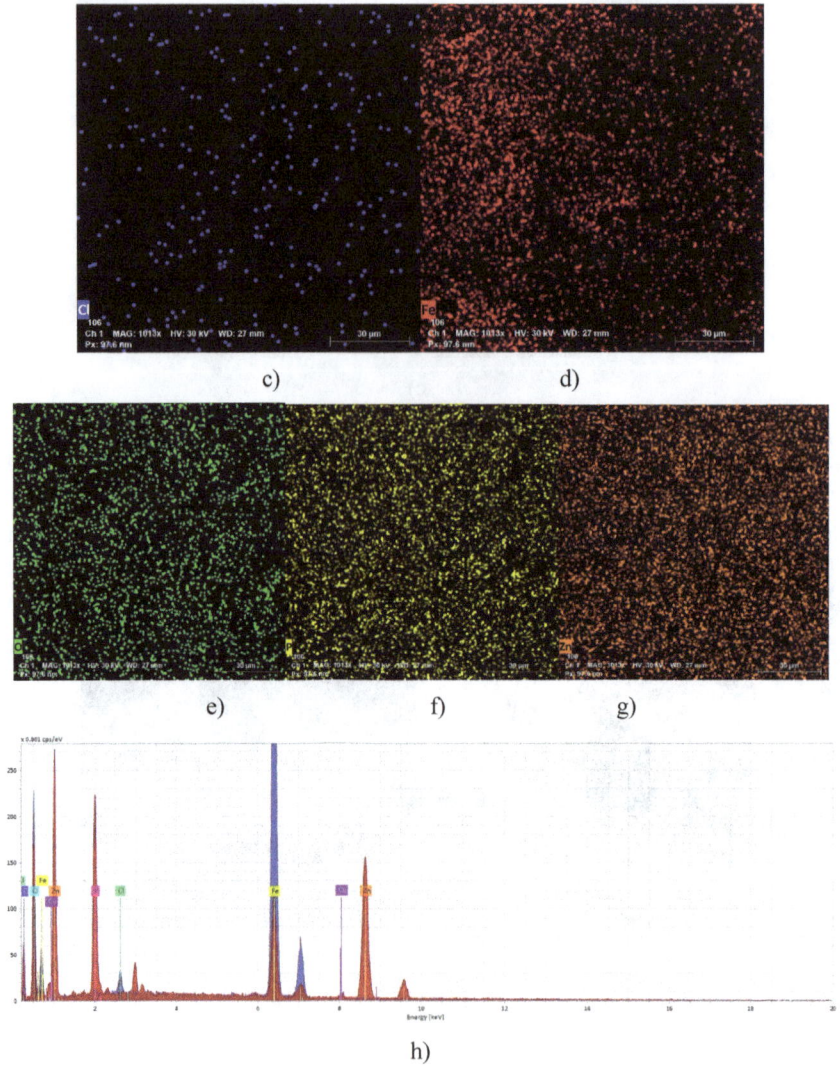

c)   d)

e)   f)   g)

h)

*Figure 5.9.   Main elements distribution on the surface of the sample phosphated with
solution II after the corrosion test a- element distribution; b- carbon distribution; c-
chlorine distribution; d- iron distribution; e- oxygen distribution; f- phosphorus
distribution; g- zinc distribution; h- energy spectrum..*

Table 5.3. *Chemical composition of OB37 steel surface after corrosion in Aegean Sea electrolyte solution.*

Element	At. No.	Mass percentages [%]	Atomic percentages [%]	EDS Error [%]
O	8	34.750	55.72	6.102
Zinc	30	32.959	12.931	0.951
P	15	13.45	11.14	0.634
Fe	26	11.88	5.298	0.466
C	6	6.8875	14.712	3.188
Cl	17	0.059	0.042	0.06

From the surface chemistry of the sample phosphated with solution II, as shown in Table 5.3, a decrease in the percentage of the chlorine element is observed, indicating that the sample was less susceptible to Cl ions and exhibited lower percentages of chloride-based corrosion compounds compared to the corrosion of OB37 steel. The high percentage of carbon can be attributed to detector errors of the EDS, and no carbon agglomerations, specific for example to surface carbonates, were observed on the distribution shown in Fig. 5.9(b).

5.3 Conclusions

It is widely acknowledged that for reinforced concrete structures, steel corrosion due to general exposure to oxygen and moisture poses the most significant threat to their stability and durability.

In the present case study, the corrosion behaviour of OB37 steel was investigated by subjecting the steel to two distinct corrosion environments: one composed of fresh water extracted from the Bahlui River and another comprising sea water drawn from the Aegean Sea.

Studies of the corrosion behaviour were performed in the two corrosion environments mentioned, both for the control sample made of steel OB37, SR 438-1:2012 and for the samples of the same steel phosphated with the three phosphating solutions presented in Chapter 2, Table 2.5 - phosphating solution I/Mg, Table 2.6 - phosphating solution II/Zn and Table 2.7 - phosphating solution III/Mn, according to the phosphating technology described in Figure 2.3.

For the OB37 control sample, the cyclic diagrams obtained in the two corrosion environments demonstrate a divergent behaviour. The diagram in freshwater from Bahlui indicates a generalized corrosion process over the entire surface of the sample, without the occurrence of passivation. Conversely, the cyclic diagram recorded in seawater exhibits characteristics of pitting, with the current density value increasing sharply as metal oxidation processes occur. At higher potentials, in parallel with metal oxidation, oxygen evolution may also occur. This phenomenon may be attributed to the elevated presence of chloride ions in the seawater.

For the OB37 sample, phosphated with I/Mg phosphating solution, the phosphating layer is enriched with Mg ions. A similar behaviour is also observed for the linear polarization of the samples in the two corrosion environments, with the anodic reaction predominating. In the context of cyclic polarization, the behaviour exhibited by the OB37 sample was found to be analogous to that of the control sample, a phenomenon that may be attributed to the high reactivity of the Mg layer. This reactivity resulted in the Mg layer piercing through, thereby allowing the electrolyte to reach the substrate.

For the OB37 sample, which was phosphated with phosphating solution II/Zn, there is a very small difference between the anodic and cathodic slopes, with the metal corroding at a much slower rate. This phenomenon can be attributed to the presence of Zn in the phosphate layer, which forms in the two corrosion media ZnO, a passivating oxide. The cyclic polarization curves, recorded in both corrosion media, are indicative of a pitting corrosion alloy, with a higher current density value in seawater due to Cl ions. It is noteworthy that beyond a certain potential, defined as the pitting potential, the metal exhibits signs of corrosion.

For the OB37 sample, which was phosphated using a III/Mn phosphating solution, no significant discrepancy between the anodic and cathodic slopes was observed, with the layer corroding at a much slower rate than for the OB37 control sample. From the cyclic polarization curves obtained, a difference in the corrosion behavior is observed in both environments, with the samples exhibiting a generalized corrosion character. This phenomenon may be attributed to the compositional complexity of the Ni, Fe, and Mn phosphate layer.

As can be seen from Table 5.1, which summarizes the corrosion process parameters obtained from the table plots for all the phosphated samples and the OB37 control sample, the best corrosion resistance was recorded for the phosphated samples with Zn solution - phosphate solution II/Zn - with the lowest corrosion rate, Vcorr, in both electrolyte media, **0.258 μm /y** for fresh water from the Bahlui and **3.060 μm /y** for sea water from the Aegean Sea. The values obtained are considerably lower than the steel corrosion rates of approximately **45 μm/y**.

The morphological changes on the surface of the samples after corrosion were evaluated comparatively by analysing the control sample of OB37 corroded in seawater and the sample of OB37 phosphated with II/Zn solution in the same electrolyte.

It was observed that the material exhibited significant damage during the corrosion test, evidenced by the deposition of corrosion compounds (iron oxides and hydroxides) on its surface. Furthermore, the elemental distribution shows the presence of chlorine and oxygen on the surface of the OB37 steel.

For the OB37 sample, which demonstrated optimal corrosion resistance, phosphated with solution II/Zn, it is observed that the coating did not suffer any major damage in the corrosion test, no corrosion marks or exfoliation were observed on the surface. The formation of micron-sized compounds on the surface during the corrosion process can also be observed. A decrease in the percentage of chlorine in the sample's surface composition indicates reduced attack by Cl ions, and the presence of chloride-based corrosion compounds is lower than in the corrosion of the OB37 steel control sample.

References

[1]   Cazac, A.-M.; Cioca, L.-I.; Lazar, P.; Badarau, G.; Cimpoesu, N.; Burduhos-Nergis, D.-P.; Iagaru, P.; Cimpoesu, R.; Cazac, A.; Bejinariu, C.; et al. Effect of Zinc, Magnesium, and Manganese Phosphate Coatings on the Corrosion Behaviour of Steel. Materials 2025, 18, 3126. https://doi.org/10.3390/ma18133126

[2]   Lazar, P.; Cimpoesu, N.; Istrate, B.; Cazac, A.M.; Burduhos-Nergis, D.-P.; Benchea, M.; Berbecaru, A.C.; Badarau, G.; Vasilescu, G.D.; Popa, M.; et al. Microstructural and mechanical properties analysis of phosphate layers deposited on steel rebars for civil constructions. Coatings 2024, 14, 182.

[3]   Creus, J.; Mazille, H.; Idrissi, H. Porosity evaluation of protective coatings onto steel, through electrochemical techniques. Surf. Coat. Technol. 2000, 130, 224–232.

[4]   Srinivasan, A.; Blawert, C.; Huang, Y.; Mendis, C.L.; Kainer, K.U.; Hort, N. Corrosion behavior of Mg–Gd–Zn based alloys in aqueous NaCl solution. J. Magnes. Alloys 2014, 2, 245–256.

[5]   Meyer, Y.A.; Menezes, I.; Bonatti, R.S.; Bortolozo, A.D.; Osório, W.R. EIS Investigation of the Corrosion Behavior of Steel Bars Embedded into Modified Concretes with Eggshell Contents. Metals 2022, 12, 417.

[6]   Zimmermann, D.; Munoz, A.G.; Schultze, J.W. Formation of Zn–Ni alloys in the phosphating of Zn layers. Surf. Coat. Technol. 2005, 197, 260–269.

[7]   Banczek, E.P.; Rodrigues, P.R.P.; Costa, I. Evaluation of porosity and discontinuities in zinc phosphate coating by means of voltametric anodic dissolution (VAD). Surf. Coat. Technol. 2009, 203, 1213–1219.

CHAPTER 6

Adhesion investigation of deposited layers

P. Lazar[1], A.-M. Cazac[1]*, C. Bejinariu[1,2]

[1]Faculty of Materials Science and Engineering, Gheorghe Asachi Technical University of Iasi, Romania

[2]Academy of Romanian Scientists, Ilfov 3, 050044 Bucharest, Romania

alin-marian.cazac@academic.tuiasi.ro

Abstract

The adhesion of the deposited layers is analysed in chapter six, the samples were embedded in concrete and then analysed on a tensile testing machine. Following the peeling process, the samples were kept in water for up to 28 days, during which time mechanical (compressive strength, splitting tensile strength) and elastic (static modulus of elasticity) determinations were performed. The determinations were according to the applicable standards: SR EN 12390-3:2019 (compressive strength), SR EN 12390-6:2023 (splitting tensile strength), SR EN 12390-13:2021 (determination of the longitudinal modulus of elasticity)

Keywords

Phosphating, Steels For Reinforced Concrete, Phosphate Layers

6.1   Mechanical properties determination of concrete

In order to determine the longitudinal modulus of elasticity in compression (i.e. Young's modulus) as well as the compressive strength, a constant loading rate of 4.7 kN/s was utilised. The splitting tensile strength was determined at a loading rate of 1.7 kN/s.

With regard to the longitudinal modulus of elasticity determination, Fig. 6.1, three loading-unloading cycles were performed, corresponding to force values of 18.06 kN and 36.1kN, as shown in Fig. 6.1. The values indicated by the digital micro-comparator (with an accuracy of 0.001 mm) were utilised to ascertain the values of the specific deformations necessary to calculate the modulus of elasticity value. In the event that the values obtained for the longitudinal modulus of elasticity differ by more than 10% across the three loading-unloading cycles, the measurement is to be repeated.

The following calculation equation (6.1) applies to the equipment used (manufactured by Humboldt AG):

$$E = \frac{2 \times \frac{F_2 - F_1}{A}}{\frac{\Delta\ell_2 - \Delta\ell_1}{100}} = 200 \times \frac{F_2 - F_1}{A(\Delta\ell_2 - \Delta\ell_1)} \quad (6.1)$$

where:

F_2 is the force value at the top of the load-unload range (in this case equal to 36.1 kN);

F_1 is the force value at the lower end of the load-unload range (in this case equal to 18.06 kN);

$\Delta\ell_2$ is the value indicated by the micro-comparator corresponding to the force at the top of the loading-unloading range (F_2);

$\Delta\ell_1$ is the value indicated by the micro-comparator corresponding to the force at the lower end of the loading-unloading range (F_1).

$A = \frac{\pi D^2}{4}$ – cross-sectional area of the cylindrical sample (D - diameter of the cylindrical sample)

100 – distance between the fixing points of the device on the cylindrical specimen (in mm)

2 – multiplier coefficient considering the leverage effect of the device

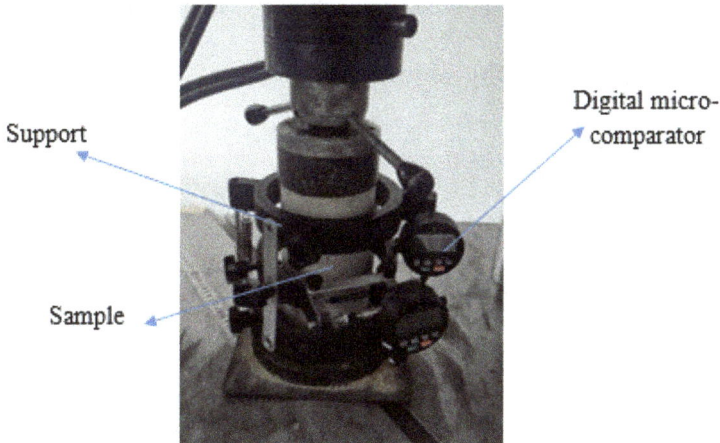

Figure 6.1.   Longitudinal modulus of elasticity determination.

The following equation was used to determine the compressive strength, Fig. 6.2:

where:

F_{max} – the maximum applied force corresponding to the failure of the cylindrical sample.\

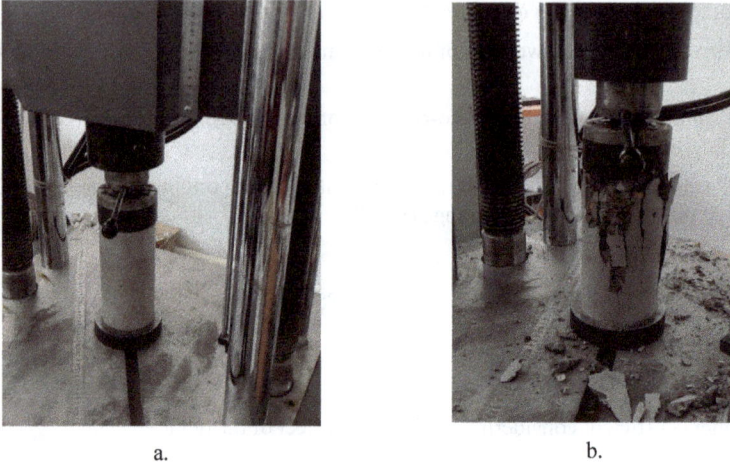

a.   b.

*Figure 6.2.   Compressive strength determination: a.- specimen before the test; b.-
specimen in failure.*

The following equation was used to determine the splitting tensile strength, Fig. 6.3:

$$f_t = \frac{2 \times F_{max}}{\pi \times D \times L} \qquad (6.3)$$

where:

F_{max} – the maximum applied force corresponding to the failure of the cylindrical sample

D – diameter of the cylindrical sample

L – length of cylindrical sample

a.   b.

Figure 6.3.   Splitting tensile strength determination: a- specimen before test; b-
specimen in failure.

Subsequent to the consolidation of the outcomes, the mechanical characteristics of the concrete utilised, ascertained at 28 days, are shown in Table 6.1.

Table 6.1.   Mechanical properties of concrete at 28 days.

Longitudinal elastic modulus [MPa]	Splitting tensile strength [MPa]	Compressive strength [MPa]
27495	4.92	38.03

As a result of the value obtained for compressive strength, it can be concluded that the concrete meets the minimum strength requirement for class C20/25.

6.2   Maximum tensile force determination and evaluation of reinforcement-concrete adhesion

The determination of the maximum tensile force at which loss of adhesion between reinforcement and steel occurs was performed in accordance with SR EN 10080:2005, Appendix D. The specimens were mounted in the test fixture, Figure 6.1, and aligned with the test equipment (Zwick Roell SP1000 universal machine, maximum force 1000 kN), thus eliminating eccentricities that could have affected the quality of the results.

According to the European Standard EN 10080:2005, Appendix D, the increase rate of the tensile force was 40 N/s for all the specimens belonging to the four cases: the control case, the untreated reinforcement and the three cases in which the reinforcement was subjected to a phosphating process with three different solutions.

The tensile force-displacement curves obtained were recorded by the equipment's acquisition system using a software program that ensures the accuracy and control of the test parameters. Furthermore, as illustrated in Fig. 6.1, a digital micro-comparator with an accuracy of 0.001 mm was utilised, positioned at the base of the specimens in contact with the steel reinforcement. This configuration enabled the identification of the moment at which the concrete reinforcement began to slip, and the observations were corroborated by the shape and values of the tensile force-displacement graphs.

6.2.1   *Maximum tensile force determination and evaluation of reinforcement-concrete adhesion*

The tensile force-displacement curves for all six specimens of the control formula are shown in Fig. 6.4. This phenomenon signifies the point at which the bond between the steel reinforcement and the concrete is lost, a phenomenon also evidenced by the sudden increase in the values indicated by the micro-comparator. This is followed by a horizontal plateau, which is associated with the effect of frictional force between the steel and the concrete. This force tends to oppose the pulling of the steel from the concrete.

Exceptions to the above are the M3 specimens, for which the force value increases even if the yield force decreases, and the M5 specimen, for which the decrease in tensile force intensity is not as sharp as for the other specimens.

The loss of adhesion between steel and concrete was also visually confirmed after the experimental tests by the appearance of a gap between the isolated part of the steel reinforcement, Figure 6.1, and the upper face of the concrete cube.

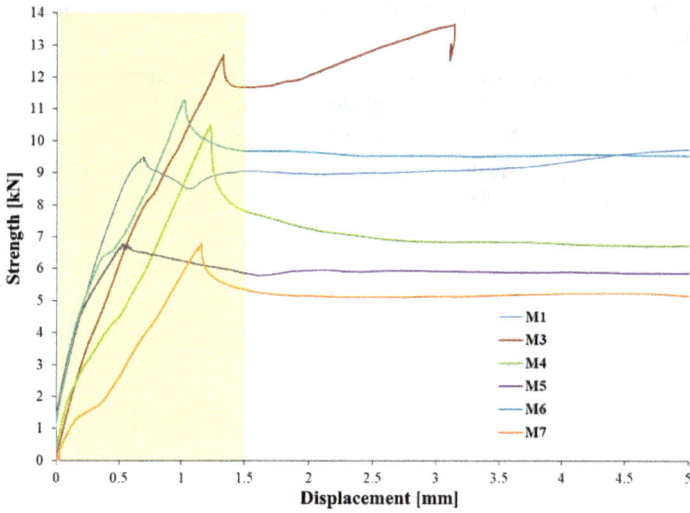

a.

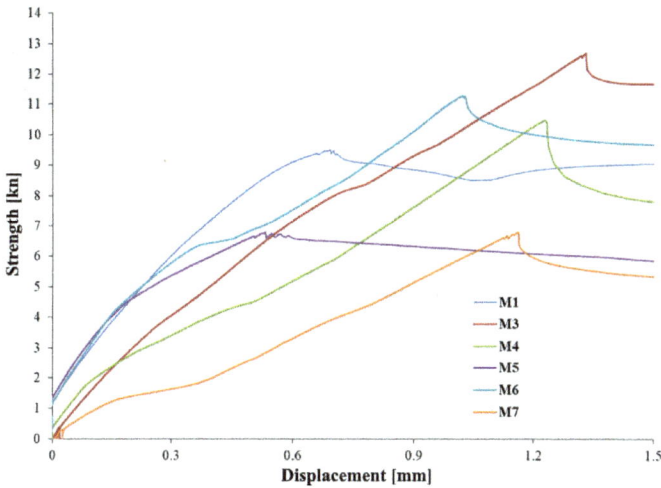

b.

Figure 6.4.   Tensile force - displacement curves for control samples: a - tensile force - displacement curves; b - detail of maximum tensile force intensity.

6.2.2 *Maximum tensile force determination for samples phosphated with solution I/Mg*

The tensile force -displacement curves for all 6 specimens corresponding to the case where the reinforcement has been phosphated with **solution I** are shown in Fig. 6.5. Conversely, the horizontal plateau of the concrete reinforcement pull-out is sustained.

In this case, the maximum tensile force at which loss of adhesion is considered to occur is that corresponding to the horizontal plane on the graph. A further comparison of the graphs presented in Fig. 6.4 and 6.5 reveals that treatment of the reinforcement with **solution I** resulted in a significant decrease in the maximum force intensity for which loss of adhesion occurs.

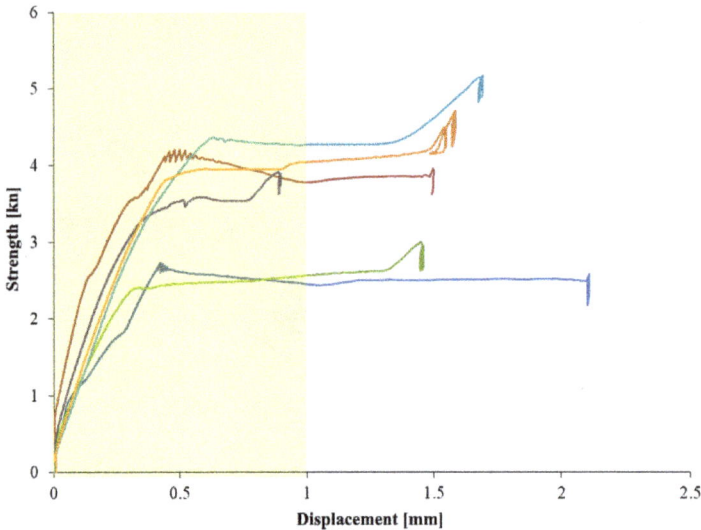

a.

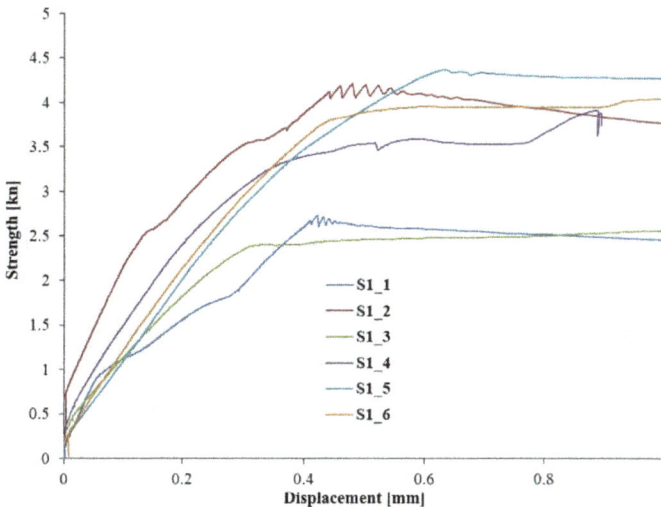

b.

*Figure 6.5.   Tensile force - displacement curves for samples treated with solution I: a-
tensile force - displacement curves; b- detail of maximum tensile force intensity.*

6.2.3   Maximum tensile force determination for samples phosphated with solution II/Zn

The tensile force-displacement curves for all 6 specimens corresponding to the case where the reinforcement was phosphated with **solution II** are shown in Fig. 6.6. In contrast to the control specimens, an increase of 13.67% in the tensile force value is observed. Furthermore, the occurrence of a precipitous decline in maximum force of adhesion, succeeded by a horizontal plateau, which was evident in the control specimens but less pronounced, reemerges. Furthermore, a discernible change in the slope of the tractive force-displacement graph is evident prior to the attainment of maximum force intensity.

A comparison of the graphs shown in Fig. 6.4 and 6.6 reveals that treatment of the reinforcement with **solution II** resulted in an increase in the maximum force intensity at which loss of adhesion occurred. It is however significantly superior to both **Solutions I** and **III** and the control specimens.

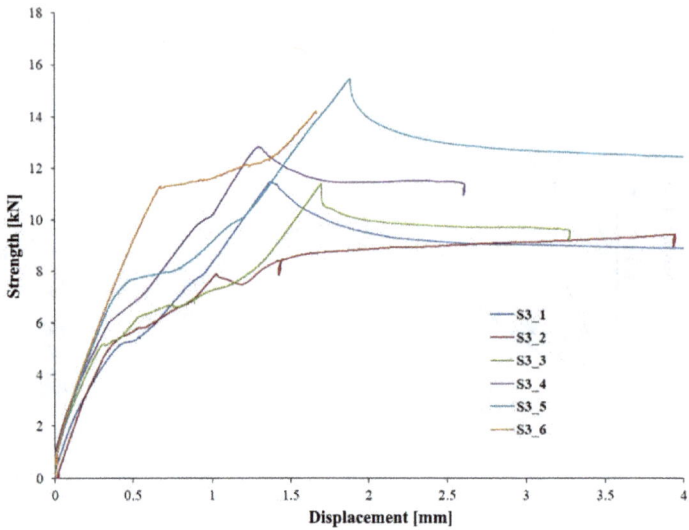

a.

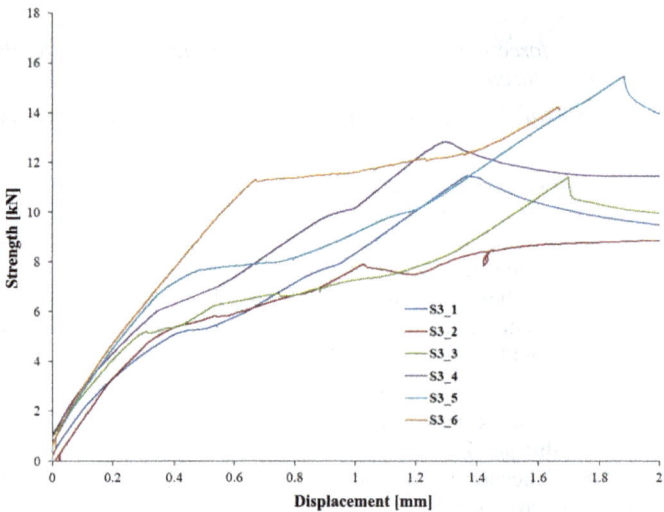

b.

*Figure 6.6.   Tensile force - displacement curves for the samples treated with solution
III: a - tensile force - displacement curves; b - detail of maximum tensile force intensity.*

6.2.4   Maximum tensile force determination for samples phosphated with solution III/Mn

The tensile force-displacement curves for 5 of the 6 specimens corresponding to the case where the reinforcement was phosphated with **solution III** are shown in Fig. 6.7. The sixth specimen exhibited a failure mode that was not consistent with the experimental tests, and was therefore excluded from the analysis. In contrast to the control specimens, a decrease of approximately 18% in the tensile force value is observed. Furthermore, the occurrence of a precipitous decline in the maximum loss of adhesion force, succeeded by a subsequent horizontal plateau, as evidenced in the control specimens, is reiterated.

A comparison of the graphs shown in Fig. 6.4 and 6.7 indicates that treatment of the reinforcement with **solution III** resulted in a decrease in the maximum force intensity at which loss of adhesion occurred. However, **solution III** was found to be significantly more effective than **solution I**, as illustrated in Fig. 6.5.

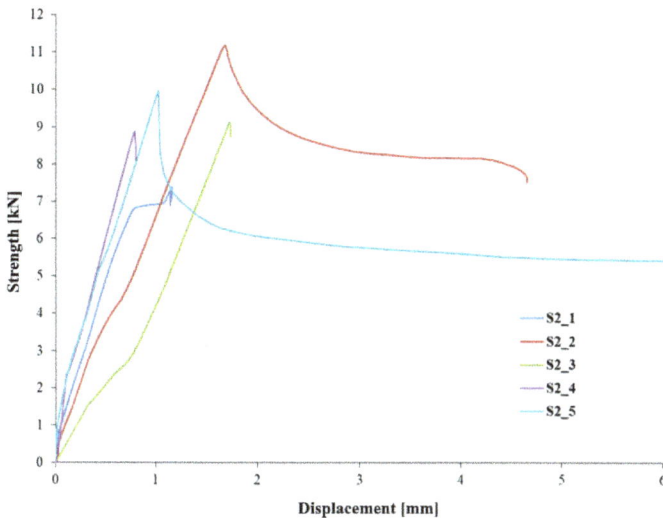

a.

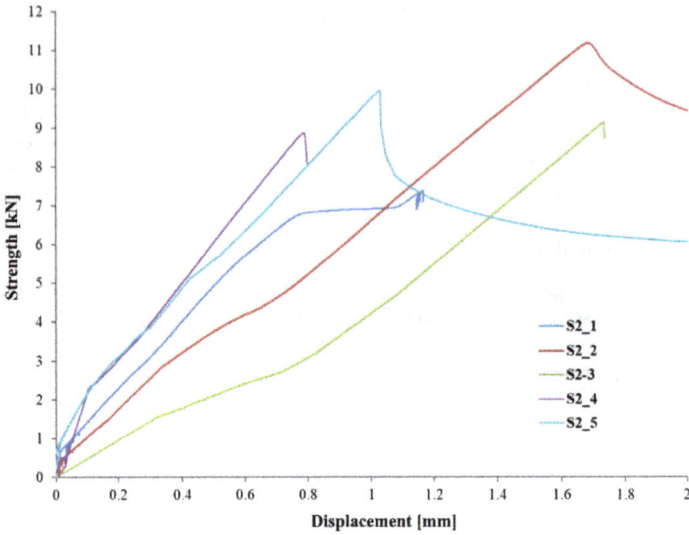

b.

*Figure 6.7.   Tensile force - displacement curves for samples treated with solution II: a-
tensile force - displacement curves; b- detail of maximum tensile force intensity.*

The results obtained are presented in Table 6.2.

Table 6.2.   Experimental results (mean values of adhesion loss force).

Samples	Fmax [kN]	Standard Deviation [-]	Coefficient of variation (COV) [%]	Failure mechanism [-]
OB37 control sample	10.97	1.353	12.34	Slipping
OB37 sample, phosphated with I/Mg solution	4.28	0.292	6.83	Slipping
OB37 sample, phosphated with II/Zn solution	12.47	1.245	9.98	Slipping
OB37 sample, phosphated with III/Mn solution	9.26	0.567	6.12	Slipping

6.3 Conclusions

From the results of the experimental results, it can be concluded that for all the specimens tested, a sliding of the reinforcement in the concrete failure mode (loss of adhesion) was obtained [1]. This failure mode is consistent with the expectations and the results previously documented in the research literature, considering the type of reinforcement used, i.e. smooth reinforcement [1,2].

A very good clustering of the results obtained can also be observed, as evidenced by coefficients of variation of less than 15%. This is indicative of the rigour with which the laboratory tests were conducted and their repeatability.

The tangential stress values at the reinforcement-concrete interface were determined on the basis of the provisions of SR EN 10080:2005, Appendix D, and the analytical relations identified in other specialized studies.

The equation proposed in SR EN 10080:2005, Appendix D is given in Equation 6.4:

$$\tau_{dm} = \frac{1}{\ell} \times \frac{F_{max}}{D^2 \frac{f_{cm}}{f_c}} \qquad (6.4)$$

where:

τ_{dm} – tangetial adhesion stress, [N/mm^2];

ℓ - the embedment length of the reinforcement in the concrete (in this case, the side of the cube), [mm];

F_{max} – the maximum tensile force at which loss of adhesion occurs, [N];

D – diameter of steel reinforcement, [mm];

f_{cm} – experimentally determined compressive strength value of concrete, [N/mm^2];

f_c – compressive strength value of the concrete considered as strength class (in this case C20/25), [N/mm^2].

The analytical form of tangential bond stress calculation proposed by SR EN 10080:2005 also includes the contribution of the concrete strength class. As demonstrated in the relevant literature, the importance of this parameter is recognised, but exactly how it can be quantified is not yet well understood [1].

Therefore, as an alternative method of evaluating the tangential stresses arising at the interface between steel and concrete and governing the bond between the two materials, the equation commonly used by researchers in this field was used [2,3]:

$$\tau_a = \frac{1}{\ell} \times \frac{F_{max}}{D^2} \qquad (6.5)$$

The results obtained are presented in Table 6.3.

Table 6.3.   The results obtained on adhesion.

Sample	F_{max} [kN]	τ_{dm} (EQ 6.4) [N/mm^2]	τ_a (EQ 6.5) [N/mm^2]
OB37 control sample	10.97	6.82	4.54
OB37 sample, phosphated with I/Mg solution	4.28	2.66	1.78
OB37 sample, phosphated with II/Zn solution	12.47	7.77	5.18
OB37 sample, phosphated with III/Mn solution	9.26	5.70	3.80

It is observed that the values of the tangential bond stress are higher when equation 4 (SR EN 10080:2005, Appendix D) is used than when the contribution of the concrete strength is neglected. While the outcomes derived from equation 5 are conservative, it is crucial to acknowledge that they could potentially result in the design of the building elements being unreasonable, with a material consumption far exceeding the requirements to achieve the safety performance criteria.

In consideration of the findings, it can be concluded that solution II for the surface treatment of rebars is a solution with great potential for application in the construction industry. Furthermore, enhancements can be made to solution III to enhance adhesion between the reinforcement and the concrete by modifying the parameters of the phosphating solution (e.g. altering the solution concentration or phosphating time).

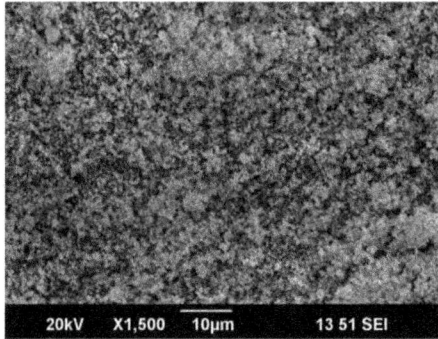

Figure 1.   Please use good resolution figures when possible. 300 dpi resolution figures are preferred. Please feel free to use colour for your figures. Both print and online version will be reproduced in full colour.

Table 1.   Sample table

Parameters	Values
a (Å)	5.5563(10)
b (Å)	7.8209(12)
c (Å)	5.5150(10)
α=β=γ (°)	90
Volume (Å^3)	239.66 (11)
Density (gm/cc)	6.70(1)

References

[1]   Shunmuga Vembu, P.R.; Ammasi, A.K. A Comprehensive Review on the Factors Affecting Bond Strength in Concrete. Buildings 2023, 13, 577. https://doi.org/10.3390/buildings13030577

[2]   Huang, H. Technology-Driven Financial Risk Management: Exploring the Benefits of Machine Learning for Non-Profit Organizations. Systems 2024, 12, 416. https://doi.org/10.3390/systems12100416

[3]   Bompa, D.V., Elghazouli, A.Y., Xu, B., Stafford, P.J., Ruiz-Teran, A.M., Experimental assessment and constitutive modelling of rubberised concrete materials, Construction and Building Materials, 137, 2017, 246-260, ISSN 0950-0618,https://doi.org/ 10.1016/j.conbuildmat.2017.01.086.

About the Authors

Petru LAZAR

Contacts: lazar.petru@gmail.com

PhD Eng. with 5 ISI papers and over 20 citations. The main contributions focused on improving the properties of steels intended for reinforced concrete by obtaining phosphated layers. I developed three phosphating solutions, the support material used was OB37 steel used in concrete reinforcement. Experimental research highlighted the improvement of the properties of OB37 steel after phosphating.

Alin-Marian CAZAC

Lecturer Ph.D. Eng.

Materials Science and Engineering Faculty,

"Gheorghe Asachi" Technical University of Iasi

Contacts: alin-marian.cazac@academic.tuiasi.ro

Lecturer at Materials Engineering and Industrial Safety Department, with 9 years of experience. He is the author of more than 48 ISI articles, Hirsh index 4 and more than 50 citations. The significant scientific contribution to most articles is the analysis of metallic materials in terms of mechanical properties. Co-author at 2 books in technical field. Member of ASESSM society. He participated as a specialist member in the research team in many scientific research contracts. Reviewer for many international Journals.

Costica BEJINARIU

Professor Ph.D. Eng.

Contacts: costica.bejinariu@academic.tuiasi.ro; costica.bejinariu@gmail.com

I am full professor and researcher at the "Gheorghe Asachi" Technical University of Iasi, Romania, with over 35 years of academic experience. I was Vice Dean of Faculty of Materials Science and Engineering (2012-2024) and I am doctoral supervisor since 2009, with seven completed doctoral thesis and seven ongoing doctoral students.

My main field of expertise is Materials Engineering: I have published a total of 31 Books/Chapters and Monographs, over 300 research papers with more than 1800 citations and a H index of 21-Web of Science / 23-Scopus / 21-ScholarGoogle. I worked at over 50 contracts of scientific research and development, in five of which I was project director and two of which project responsible. Furthermore, I own 12 invention patents in the field of Materials Engineering, which were presented at numerous Invention Salons, winning medals.

I am an active member of professional associations in the field, reviewer for scientific journals and in scientific committees at different congresses. I am a member in the Committee of Materials Engineering, domain Engineering and Materials Science of the National Council for Attestation of University Titles, Diplomas and Certificates (CNATDCU), which is an independent consulting entity at a national level in Romania.

I am an Associate Member of the Academy of Romanian Scientists.